Philosophy and Engineering Education

New Perspectives, An Introduction

Synthesis Lectures on Engineering, Science, and Technology

Each book in the series is written by a well known expert in the field. Most titles cover subjects such as professional development, education, and study skills, as well as basic introductory undergraduate material and other topics appropriate for a broader and less technical audience. In addition, the series includes several titles written on very specific topics not covered elsewhere in the Synthesis Digital Library.

Integrated Process Design and Operational Optimization via Multiparametric Programming
Baris Burnak, Nikolaos A. Diangelakis, and Efstratios N. Pistikopoulos
2020

The Art of Teaching Physics with Ancient Chinese Science and Technology
Matt Marone
2020

Scientific Analysis of Cultural Heritage Objects
Michael Wiescher and Khachatur Manukyan
2020

Case Studies in Forensic Physics
Gregory A. DiLisi and Richard A. Rarick
2020

An Introduction to Numerical Methods for the Physical Sciences
Colm T. Whelan
2020

Nanotechnology Past and Present
Deb Newberry
2020

Introduction to Engineering Research
Wendy C. Crone
2020

Theory of Electromagnetic Beams
John Lekner
2020

The Search for the Absolute: How Magic Became Science
Jeffrey H. Williams
2020

The Big Picture: The Universe in Five S.T.E.P.S.
John Beaver
2020

Relativistic Classical Mechanics and Electrodynamics
Martin Land and Lawrence P. Horwitz
2019

Generating Functions in Engineering and the Applied Sciences
Rajan Chattamvelli and Ramalingam Shanmugam
2019

Transformative Teaching: A Collection of Stories of Engineering Faculty's Pedagogical Journeys
Nadia Kellam, Brooke Coley, and Audrey Boklage
2019

Ancient Hindu Science: Its Transmission and Impact on World Cultures
Alok Kumar
2019

Value Rational Engineering
Shuichi Fukuda
2018

Strategic Cost Fundamentals: for Designers, Engineers, Technologists, Estimators, Project Managers, and Financial Analysts
Robert C. Creese
2018

Concise Introduction to Cement Chemistry and Manufacturing
Tadele Assefa Aragaw
2018

Data Mining and Market Intelligence: Implications for Decision Making
Mustapha Akinkunmi
2018

Empowering Professional Teaching in Engineering: Sustaining the Scholarship of Teaching
John Heywood
2018

The Human Side of Engineering
John Heywood
2017

Geometric Programming for Design Equation Development and Cost/Profit Optimization (with illustrative case study problems and solutions), Third Edition
Robert C. Creese
2016

Engineering Principles in Everyday Life for Non-Engineers
Saeed Benjamin Niku
2016

A, B, See... in 3D: A Workbook to Improve 3-D Visualization Skills
Dan G. Dimitriu
2015

The Captains of Energy: Systems Dynamics from an Energy Perspective
Vincent C. Prantil and Timothy Decker
2015

Lying by Approximation: The Truth about Finite Element Analysis
Vincent C. Prantil, Christopher Papadopoulos, and Paul D. Gessler
2013

Simplified Models for Assessing Heat and Mass Transfer in Evaporative Towers
Alessandra De Angelis, Onorio Saro, Giulio Lorenzini, Stefano D'Elia, and Marco Medici
2013

The Engineering Design Challenge: A Creative Process
Charles W. Dolan
2013

The Making of Green Engineers: Sustainable Development and the Hybrid Imagination
Andrew Jamison
2013

Crafting Your Research Future: A Guide to Successful Master's and Ph.D. Degrees in Science & Engineering
Charles X. Ling and Qiang Yang
2012

Fundamentals of Engineering Economics and Decision Analysis
David L. Whitman and Ronald E. Terry
2012

A Little Book on Teaching: A Beginner's Guide for Educators of Engineering and Applied Science
Steven F. Barrett
2012

Engineering Thermodynamics and 21st Century Energy Problems: A Textbook Companion for Student Engagement
Donna Riley
2011

MATLAB for Engineering and the Life Sciences
Joseph V. Tranquillo
2011

Systems Engineering: Building Successful Systems
Howard Eisner
2011

Fin Shape Thermal Optimization Using Bejan's Constructal Theory
Giulio Lorenzini, Simone Moretti, and Alessandra Conti
2011

Geometric Programming for Design and Cost Optimization (with illustrative case study problems and solutions), Second Edition
Robert C. Creese
2010

Survive and Thrive: A Guide for Untenured Faculty
Wendy C. Crone
2010

Geometric Programming for Design and Cost Optimization (with Illustrative Case Study Problems and Solutions)
Robert C. Creese
2009

Style and Ethics of Communication in Science and Engineering
Jay D. Humphrey and Jeffrey W. Holmes
2008

Introduction to Engineering: A Starter's Guide with Hands-On Analog Multimedia Explorations
Lina J. Karam and Naji Mounsef
2008

Introduction to Engineering: A Starter's Guide with Hands-On Digital Multimedia and Robotics Explorations
Lina J. Karam and Naji Mounsef
2008

CAD/CAM of Sculptured Surfaces on Multi-Axis NC Machine: The DG/K-Based Approach
Stephen P. Radzevich
2008

Tensor Properties of Solids, Part Two: Transport Properties of Solids
Richard F. Tinder
2007

Tensor Properties of Solids, Part One: Equilibrium Tensor Properties of Solids
Richard F. Tinder
2007

Essentials of Applied Mathematics for Scientists and Engineers
Robert G. Watts
2007

Project Management for Engineering Design
Charles Lessard and Joseph Lessard
2007

Relativistic Flight Mechanics and Space Travel
Richard F. Tinder
2006

Philosophy and Engineering Education: New Perspectives, An Introduction
John Heywood, William Grimson, Jerry W. Gravander, Gregory Bassett, and John Krupczak, Jr.

ISBN: 978-3-031-03751-1   paperback
ISBN: 978-3-031-03761-0   PDF
ISBN: 978-3-031-03771-9   hardcover

DOI 10.1007/978-3-031-03761-0

A Publication in the Springer series
SYNTHESIS LECTURES ON ENGINEERING, SCIENCE, AND TECHNOLOGY

Lecture #19
Series ISSN
Print 2690-0300   Electronic 2690-0327

Philosophy and Engineering Education

New Perspectives, An Introduction

John Heywood
Trinity College Dublin–University of Dublin

William Grimson
Dublin University of Technology

Jerry W. Gravander
Clarkson University

Gregory Bassett
Hope College, Michigan

John Krupczak, Jr.
Hope College, Michigan

SYNTHESIS LECTURES ON ENGINEERING, SCIENCE, AND TECHNOLOGY #19

ABSTRACT

All educators bring to their work preconceived ideas of what the curriculum should be and how students learn. Seldom are they thought through. Since without an adequate philosophical base it is difficult to bring about desirable changes in policy and practice, it is necessary that educators have defensible philosophies of engineering education. This point is illustrated by recent debates on educational outcomes which can be analysed in terms of competing curriculum ideologies.

While these ideologies inform the development of a philosophy of engineering education they do so in light of a philosophy of engineering for such a philosophy focuses on what engineering is, and in particular how it differs from science. This is addressed in this study through consideration of the differences in the modes of abstraction required for the pursuit of science on the one hand, and the pursuit of engineering design, on the other hand.

It is shown that a philosophy of engineering is not a philosophy of science or a philosophy of engineering education, but it is from a philosophy of engineering that a philosophy of engineering education is drawn. Uncertainty is shown to be a key characteristic of engineering practice.

A way of formulating a philosophy of engineering is to consider it through the classical prism that splits the subject into five divisions, namely epistemology, metaphysics, logic, ethics aesthetics. Additionally, "behaviour" also characterizes the practice of engineering.

KEYWORDS

abstract thinking, aesthetics, behavior, curriculum ideologies, constructivism, design (nature of design ideas-theory of), engineering function, engineering method, engineering practice, epistemology, ethics, logic, metaphysics, philosophy (-of engineering, -of engineering education, -of science), realism, science, uncertainty

Contents

Series Forword xiii

Preface xv

Acknowledgments xvii

1   Philosophy and Engineering Education: Should Teachers Have a
     Philosophy of Education? 1
     John Heywood
     1.1   Abstract 1
     1.2   Beyond Operational (Working) Philosophy to a Defensible Theory of
           Engineering Education 1
     1.3   The ABET Debate 4
     1.4   The Technical Dimension: Language and the Meaning of Things 4
     1.5   From the Outside Looking In 5
     1.6   Ideologies Behind the Debate 5
     1.7   Discussion 10
     1.8   Notes and References 12

2   Engineering and Philosophy 17
     William Grimson
     2.1   Abstract 17
     2.2   Introduction 17
     2.3   The Five Main Branches of Philosophy (Classical) 20
     2.4   Behaviour 23
     2.5   Conclusion 26
     2.6   Notes and References 27

3   Philosophy of Engineering as Propaedeutic for the Philosophy of
     Engineering Education 31
     Jerry W. Gravander
     3.1   Abstract 31

3.2   Introduction 31

3.3   The Philosophy of Engineering Is Not the Philosophy of Science 32

3.4   The Philosophy of Engineering Narrowly Defined 33

3.5   Philosophy of Engineering Is Not Philosophy of Engineering Education . . . 33

3.6   Engineering Uncertainty 34

3.7   Implications of Engineering Uncertainty for the Philosophy of
       Engineering Education 36

3.8   Broadening the Argument 37

3.9   Conclusion 38

3.10   Notes and References 38

4   Abstract Thought in Engineering Science: Theory and Design **41**

Gregory Bassett and John Krupczak, Jr.

4.1   Abstract 41

4.2   Introduction 41

4.3   The Standard Account 42

4.4   Products 43

4.5   Methods 45

4.6   Education 48

4.7   Notes and References 49

Authors' Biographies **53**

Series Foreword

In 2011 The Educational Research and Methods division of the American Society for Engineering Education (ASEE), The Education Society of the Institute of Electrical and Electronic Engineers, and the National Science Foundation sponsored a one-day workshop at the annual Frontiers in Education Conference (FIE) on "Exploring the Philosophies of Engineering and Engineering Education". The workshop arose from the evident interest in philosophical issues demonstrated by attendance at a series of workshops, papers, and special sessions held at the annual Frontiers in Education Conferences between 2007 and 2010.

Sue Kemnitzer of the National Science Foundation believed that every engineering educator should have a philosophy of engineering, a view that was shared by workshops sponsoring official in the NSF Dr. Alan Cheville. They believed that by having a philosophy of engineering education it would serve as a step towards theory development in engineering education that would go beyond simply adapting more general learning theories given that engineering epistemologies are not necessarily aligned with those of science and mathematics. They hoped, therefore, that a way could be found to continue the work of the workshop and its antecedents. In the event, a formal home for philosophy was found in the Technological Literacy Division of ASEE. This came at a time when the Division was considering its role in relation to engineering literacy. Thus, in 2013 The Board of ASEE agreed to a change in name and the objectives of the Division. It became the Technological and Engineering Literacy and Philosophy of Engineering Division (TELPhE), and took on the role wished for by Sue Kemnitzer. FIE continues to hold paper sessions and workshops devoted to the subject. TELPhE has as its goal the development of innovative curricula and delivery methods for the assessment of technological and engineering literacy education. Since an understanding of engineering is a critical element of technological literacy, the division supports efforts to develop a philosophy of engineering and technology. The Division encourages collaboration between people with engineering backgrounds and people with backgrounds outside of engineering, as well as with cognate divisions in ASEE.

All of us have operational philosophies that drive our actions at work, in the community and in the family. Few of us think them through and for this reason the goals of engineering education remain as goals. While introducing the philosophies of engineering and education this series asks the reader to question their operational philosophies with a view to actively making them more coherent, more robust, and more applicable and useful to address the needs of present and future students, industry and society.

This series of three texts has been developed from work undertaken by members of TELPhE and their associates who have become interested in philosophy more generally, and particular philosophers more specifically (e.g. Dewey, James, Lonergan, Macmurray, Newman, Pierce, Whitehead), who like Wittgenstein, view *philosophy not as a theory but an activity*.

November 2021

Preface

The overarching aim of this series of texts is to illustrate the importance of the engineering educator's personal philosophies of engineering and engineering education to his/her practice. Indeed, it has been argued that the failure to achieve the goals of engineering education is due to the fact the engineering education lacks a philosophical base [1].

Viewed from one perspective it is argued that a philosophy of engineering education is not possible without a philosophy of engineering. From the perspective of the goals of engineering education that would seem to be self-evident. In this introductory text this point is demonstrated by Jerry Gravander in Chapter 3. Prior to that, in Chapter 2 William Grimson gives an account of how philosophy is relevant to engineering by referencing five classical branches of philosophy.

Engineers need a philosophy to answer such questions as "How does engineering relate to science"? In recent years, many papers have explored the issue posed by this question. In Chapter 4 of this volume, Gregory Bassett and John Krupczak suggest that the differences lie in the type of abstraction that is used to answer the questions peculiar to engineering on the one hand, and questions that are peculiar to science, on the other hand.

It will be evident to the reader that currently engineering is raising profound philosophical questions for engineers. For example, "What degree of responsibility should engineers take in relation to the social impact of works of engineering"? While it is not the purpose of these texts to discuss these issues it is its goal to invite the reader into the world of philosophy and philosophers that have stimulated the authors of the various chapters in order in order that the reader can begin "to learn to be aware of problems in your thinking where you might not have suspected them", for that according to Jonathan Rée is what philosophy is [2].

There is, however, another perspective. It is of the engineering educator as instructor. There is little or no escape for most engineering educators from teaching. That is what they are called upon to do from day one. In this respect, they are no different to school teachers. All of us enter teaching with an operational philosophy, that is, with a set of beliefs about how students learn. These beliefs drive our motivation, and we seldom question them. Chapter 1 presents four different ideologies common to all levels of education that drive the beliefs of individuals in the hope that in the activity of thinking about them the reader will renew, develop, or change his/her stance and so enhance the learning of his/her students.

NOTES AND REFERENCES

[1] Sinclair, G. and Tilston, W. (1979). Improved goals for engineering education. *Proceedings ASEE/IEEE Frontiers in Education Conference*, 3, A 25–31. xv

[2] Cited by William Grimson, Chapter 2. xv

Acknowledgments

The chapters in this book, with the exception of the first, are based on adaptations of papers that originally appeared in *Philosophical and Educational Perspectives in Engineering and Technological Literacy* Handbooks 1 Published by Original Writing, Dublin on behalf of the Technological and Engineering Literacy and Philosophy Division of the American Society for Engineering Education, Washington DC. They may be retrieved at http://lib.dr.iastate.edu/ece_books/2, ISBN Parent 9781-1-78237-567-8.

NOTES AND REFERENCES

[1] Grimson, W. *Engineering and Philosophy*, pp. 26–36.

[2] Gravander, J. W. *Philosophy of Engineering as Propaedeutic for the Philosophy of Engineering Education*, pp. 37–46.

[3] Bassett, G. and Krupczak, J., Jr., *Abstract Thought in Engineering and Science: Theory and Design*, pp. 47–57.
Chapter 1 is an adaptation of a paper with the same title given at the 2017 annual conference of the American Society for Engineering Education and is reprinted here with the permission of the Society.

November 2021

CHAPTER 1

Philosophy and Engineering Education: Should Teachers Have a Philosophy of Education?

John Heywood

1.1 ABSTRACT

Arguments for engineering educators having a formulated philosophy of engineering education are presented. The perspectives that a person takes to moral dilemmas will be driven by the beliefs they have about the nature of morality and truth. Similarly, the beliefs that a person has about the content of the curriculum, instruction, and learning will, in all likelihood, be founded on one of the great "isms" of philosophy. A discussion of the debate that followed the publication of the recent revision of the ABET criteria in light of four curriculum ideologies is used to illustrate this argument. As would be expected, such conference debates are conducted at a "surface" level when they need to be conducted "in depth." But this cannot happen without every engineering educator being versed in the philosophies that these different positions embrace. It is doubtful if hard and fast positions would then be maintained, and it would be strange if there was no renewal of the curriculum with an attendant restructuring. Such is the need for professional training in pedagogy in which educators are given the opportunity to explore a variety of philosophies and learning approaches. Such activity is philosophy, and the result will be an articulated and defensible philosophy of engineering education.

1.2 BEYOND OPERATIONAL (WORKING) PHILOSOPHY TO A DEFENSIBLE THEORY OF ENGINEERING EDUCATION

It comes as a shock to many people to find their thoughts are not as independent as they feel them to be. They find that many of the attitudes they possess are formed not independently but by many external forces that impinge upon them. It comes as a powerful shock to find that the family, and more than that, the interactions with their peer groups at school and work have had a lasting impact on their behaviour. They would no doubt understand that the purpose of schooling was to impact on them. Taken together they may wonder what independence of thought they have. Fortunately, most of us avoid the trauma that such thoughts bring and carry on as if the things we do are driven totally by the activity of our free will. In contrast, we often want to be able to influence others, and in some cases we are lucky enough to have that as a job as, for example, teachers, therefore, by definition engineering educators.

To put it in another way, the operational (working) philosophies that drive our belief systems and consequently our behaviours do not arise from independent thinking but from the way in which we interact with other persons [1]. Indeed, as the philosopher John Macmurray concluded that we come to know who we are as individuals only in personal relationships [2]. But we do make decisions and it is in the making of the decisions that we begin the act of learning. For the most part, most of us do not question the operational philosophies or belief systems that drive our learning. In any case, for the most part, they are implicitly held. Consider, for example, how our philosophy of engineering, that is, what we believe engineering to be, was formed. Was it simply a form of applied science or something more? Then consider, if after experience in industry, this view has changed. Now, if we are considering teaching or are teaching, should we spend some time considering what engineering is? The volumes in this series are intended to present the reader with this challenge.

It is argued that every engineering educator should have a thoughtout view of what engineering is, because what happens in engineering should dictate in no small way the process of the curriculum, its goals, and the means of achieving them. By curriculum, it means all the formal and informal factors (e.g., organizational structure, peers, tutorial arrangements) that influence the motivation to learn [3–6].

Many argue, as do William Grimson and Jerry Gravander in Chapters 2 and 3, that a philosophy of engineering education that underpins what we do in practice is not possible without first developing our operational philosophies of engineering beyond the implicit to something substantial and explicit. And that, argues William Grimson in Chapter 2, may be achieved by an understanding of the "isms" of philosophy as they have developed from Aristotle and Plato onwards. To put it in another way, engineers need to understand who they are if they are to provide a curriculum that is to serve the needs of engineering. That this is important is evident from the large volume of literature that has emerged on the differences between engineering and science [7], a topic that is considered by Gregory Bassett and John Krupczak in Chapter 4.

The least that can be said is that by definition an engineer is a change agent. It follows that the results of what engineers do will necessarily require changes in the curriculum. Yet it has proved difficult to do and, therefore, the goals that many engineers and engineering educators think desirable have not been obtained [8]. A strong case has been made that the reason for this is that engineering educators have lacked a proper philosophical base that would provide the guidelines required [9].

An alternative position arises from the view that since engineering education is simply the application of science (mainly physics) to the solution of practical problems, its philosophical bases are to be found in the philosophy of science education for which a significant literature exists, e.g., [10, 11]. Indeed many science and engineering educators are committed to a constructivist perspective of learning about which much has been written in science [12]. Yet, whilst there is a strong case for examining that literature with respect to teaching the applied sciences, there is an equally substantial literature that shows that engineering is something more than the application of science to the solution of problems, not least in the way designs are born, developed, and implemented as Bassett and Krupczak show in Chapter 4. They echo the view expressed in several substantive texts that engineering is a different way of thinking to that of the scientist [13, 14]. Gravander is adamant in Chapter 3 that philosophy of engineering is not philosophy of science.

Within the practice of engineering, different philosophies lead to different perspectives on ethical issues [15] that have a major bearing on the curriculum offered to students. For example, "realists" take a "correspondence" theory of truth; that is, a statement is true if it corresponds to a state of affairs independent of the statement. In contrast, constructivists who are also relativists (they need not be relativists) take a "coherence" theory of truth; that is, a statement is true if it coheres or fits with other statements that are true. "Truth, they will claim, is constructed by human beings within the societies in which they live. In morality, therefore, there is no search for any reality beyond the moral rules human beings create and live by" [16]. Realists, on the other hand, argue that there is only one set of truths and that the task of moral philosophy is the search for those truths. To defend a moral position one has to be sure about the basis of one's beliefs; that is, they have to be defensible.

Exactly the same applies to reasoning about educational issues which often does not extend much beyond the trivial when compared to the knowledge that is available. The recent debate about the proposed revision of the ABET criteria falls into this category, and illustrates the failure of the engineering fraternity to have a fundamental debate about the aims of engineering education that is other than a set of warring opinions. But if the participants in any debate have not understood the philosophical basis of their opinions that is to be expected. Prior knowledge is a prerequisite to understanding any issue including knowledge of one's own assumptions and predispositions. Hence the need for engineering educators to have an articulated and defensible philosophy of education when discussing the goals, content, and pedagogy of engineering education. It is necessary, therefore, that in providing the base for a philosophy of engineering education it is informed by philosophies of education. These points will be illustrated by reference to the debate about ABET's proposals to change their criteria.

1.3   THE ABET DEBATE

Two dimensions of this debate will be considered. The first, for want of a better term, is called "technical." It is about the design of the criteria and need to attend to "meaning." It justifies attention to the precepts of analytic philosophy and the meaning of statements. The second dimension might best be described by the term "philosophical." It seeks to understand the belief systems that drive the curriculum debate, for it is about what content the new regulations allow and what content they do not.

1.4   THE TECHNICAL DIMENSION: LANGUAGE AND THE MEANING OF THINGS

Although the average member of the public, and for the most part that is you and I, would not want to engage in the abstract conversations of philosophers on language, some things have trickled down into the public arena. For example, the analytic philosophers of the 20th century have made us increasingly aware of the need to clarify meaning: we know that if the questions we set in a public examination are unclear there is the possibility that we will be taken to court. More pertinently, we know that if an instruction we give to a technician is misunderstood, and leads to an accident, that we are ultimately responsible for what happened. So we need to check that our instructions are understood and not misunderstood.

Nowhere does the problem of meaning raise its ugly head more than in the interpretation of statistics, particularly those to be found in newspapers, e.g., on forecasts of the state of the economy. Since the year 2000, engineering educators in the U.S. have been required by ABET to ensure that the programs they teach will achieve certain specified outcomes. Before they were introduced in the year 2000, engineering educators were able to attend meetings that clarified the meaning of these outcomes. Two engineering educators, Yokomoto and Bostwick, argued that "secondary meanings of some words are sometimes used, such as using the term 'criteria' to describe the level of performance that students must achieve and 'outcomes' to describe the learning behaviours students must demonstrate" [17]. A more common definition of "outcome" is "result" or "consequence," and anyone attaching that meaning to the word will surely become confused in any discussion about writing measurable outcomes. Yokomoto and Bostwick said that the aims listed by ABET were considered to be too broad to be assessed directly, and in the tradition of *The Taxonomy of Educational Objectives* they recommended that those aims should be broken down into smaller, more measurable units [18]. The essence of their argument was that accrediting agencies should explain the terms used, and use them consistently, and to this end they made a distinction between course outcomes and course instructional objectives. Again, such distinctions are debatable.

More generally, an important aspect of language is its use in the expression of the emotions. One effect of the outcomes movement is that it has removed many words from the language of the academic common room. One term that has many meanings and is not easy to define is "motivation" yet, it is very much a driver of our teaching—the desire to motivate both students and teachers. It is very much the language of the psychology of learning—readiness to learn, reinforcement, transfer of learning, critical thinking, problem solving, and so on. The ABET debate certainly generated many emotions.

1.5   FROM THE OUTSIDE LOOKING IN

While ABET is an American organization, in recent years its influence has extended beyond its borders and some countries are using its accreditation mechanisms. There is, therefore, international interest in the basis of the accreditation criteria which is the justification for an outsider like me, even though I am a member of ASEE using ABET, to exemplify the central thesis of this chapter, namely that every teacher should have a defensible philosophy of education. For this reason, an American paradigm developed by Michael Schiro [19] which reflects developments in school education in the U.S., and the philosophies that have driven them, is used as a focus for the argument. Schiro distinguishes between scholar academic, social efficiency, learner-centred, and social reconstruction ideologies. I argue that the ABET debate, as I was able to observe it, was a conflict between different ideologies.

1.6   IDEOLOGIES BEHIND THE DEBATE

(1) **The scholar academic ideology**—John Eggleston, an English educational sociologist and technical educator, has described a "received" paradigm of the curriculum which helps to introduce Schiro's scholar academic ideology [20]. Knowledge in this curriculum paradigm is received and accepted as given. It is non-negotiable, non-dialectic, and consensual. Knowledge is something that is given and, consequently, is that which should be transmitted to students. Through it the accumulated wisdom of a culture is transmitted. Eggleston's paradigm is similar to the "scholar academic ideology" proposed by Schiro. "Scholar academics" writes Schiro, "assume that the academic disciplines, the world of the intellect, and the world of knowledge are loosely equivalent. The central task of education is taken to be the extension of the components of this equivalence, both on the cultural level as reflected in the discovery of new truth, and on the individual level, as reflected in the enculturation of individuals into civilization's accumulated knowledge and ways of knowing" [21].

Jerome Bruner, a distinguished American psychologist, wrote: "A body of knowledge enshrined in a university faculty and embodied in a series of authoritative volumes is the result of much prior intellectual activity. To instruct someone in these disciplines is not a matter of getting him to commit results to mind. Rather it is to teach him to participate in the process that makes possible the establishment of knowledge. We teach a subject not to produce little living libraries on that subject, but rather to get a student to think mathematically for himself, to consider matters as historian does, to take part in the process of knowledge-getting. Knowing is a process, not a product" [22].

The process that makes possible the establishment of knowledge is, in this ideology, what is understood by learning. For each school subject there must be a corresponding academic discipline as represented in the universities. Because the disciplines are dynamic they are concerned as much with "what will be" as with "what was" [23]. That this is so, is illustrated by the great curriculum projects that were undertaken in the 1960s and 1970s because in the U.S. teachers did not have the resources to undertake such developments which normally are considered to be part of the role of the teacher functioning in this ideology [24].

The scholar academic ideology is teacher centred. Information is conveyed to the mind which reasons about it, as required. Learning is the result of teaching [25]. Because each discipline has within it, its own theory of learning, generalized theories of learning have no place in the design of instruction. It is not unreasonable to suggest that the majority of engineering educators would hold this ideology to be true. However, they have had to accept modifications to meet the requirements of accreditation authorities, sometimes prompted by politicians who are motivated by the "social efficiency ideology."

(2) **The social efficiency ideology** requires that the curriculum serves utilitarian purposes, namely the creation of wealth. Institutions have to be run like businesses: therefore, the curriculum has to be seen to be providing measurable outcomes in the form of objectives now called outcomes. In this paradigm the teacher's role is to guide (manage, direct, and supervise) the learner to achieve the outcomes (or terminal performances) required. Knowledge is defined behaviourally in terms of what a student "will be able to do," as a result of learning. There is little concern for the student except for the potential they have as graduates, and the inputs they give to the economy.

Evaluation and assessment are central to the vison of this ideology. It is the prevailing curriculum ideology in engineering education, as seen for example in the current ABET criteria. The social efficiency ideology has its origins in the objectives movement and the curriculum model of Ralph Tyler [26]. But, Schiro also considers that educators who subscribe to this ideology value a programmed curriculum, and the psychology underpinning it to be found in behavioural psychology, as for example that of B. F. Skinner. In engineering education it can be seen in the systems of mastery learning and personalised instruction that were experimented with in the 1960s and 1970s [27–29].

While behavioural psychology was replaced by cognitive psychology it is relevant to note that there are many politicians and administrators who believe that computer-assisted learning might come to be used to replace lectures which they considered to be conveyors of the same knowledge that is to be found in textbooks. Evaluation is very important to those who hold this ideology. There are tensions between those who adhere to the scholar academic ideology as well as those who adhere to the learning-centred and social reconstruction ideologies and the social efficiency ideology.

(3) **The learning-centred ideology** is in stark contrast to the social efficiency ideology. The child is at the centre of, and has a profound influence on, the curriculum process. This ideology is associated with the educational philosophy of John Dewey. A major feature of his approach is inquiry based learning (see Chapter 2 by Mani Mina in Volume 2 of this series) Learning-centred schools like the Montessori schools will organized in a totally different way to traditional schools.

Learner-centred schools are based on natural developmental growth rather than on demands external to them. "Individuals grow and learn intellectually, socially, emotionally and physically in their own unique and idiosyncratic ways and at their own individual rates rather than at a uniform manner" [19, p. 111]. The philosophy that underpins these schools is constructivism. The schools and curriculum are designed to produce students who are "self-activated makers of meaning, as actively self-propelled agents of their own growth, and not as passive organisms to be filled or moulded by agents outside themselves" [30]. Learning moves from the concrete to the abstract. The idea of active learning has become part of the vocabulary of higher education, not as yet in the sense of organizing an institution for active learning, but in the sense of teachers organizing and managing their classrooms such learning. The relationship between the teacher and the student is quite different to those between students and educators who follow the scholar academic or social efficiency ideologies, and Cowan [31] argues, to be preferred.

In sum, the core theses of constructivism are:

1. Knowledge is actively constructed by the cognizing subject not passively received from the environment.

2. Coming to know is an adaptive process that organizes one's experiential world; it does not discover an independent, pre-existing world outside of the mind of the knower [32].

The laboratory has been found to be a good place to apply constructivist principles in engineering [33]. The project method seems to have been first introduced these schools (see Chapter 2 by Mani Mina in Volume 2 of this series). Problem-based learning was practiced in medicine first, and then engineering is in the tradition of this ideology [34].

The idea of negotiating the curriculum has its origins in the constructivist approach [35]. Given that the reality we have is the result of our environment then, in these circumstances, the students with their teachers should design a curriculum that is real to them. In this sense, the curriculum should be negotiable and worked out to suit the individual needs of students. This is the principle behind the "independent study degrees" that have been offered in the UK [36]. In an Engineering Science University entry-level examination in the UK, students negotiated the projects they were required to undertake with their teachers and the examiners [37]. A key feature of inquiry-based learning and, therefore, of project work is the need to reflect on what has been achieved. Educators in higher education have taken on board the idea of learning how to learn or metacognition as understanding how we learn is now called.

In addition to establishing the environment for learning, the teacher has the functions of observing and diagnosing individual needs and interests, and facilitating the growth of the students in their care. Learning-centred educators are opposed to the psychometric testing a required by social efficiency educators. Standardized tests are anathema to learner-centred educators. It is believed that students' work should be assessed by the students themselves through learning logs and journals (portfolios). Some engineering educators are advocates of peer and self assessment as well as the use of portfolios and journals [38–41].

Because knowledge is created by individuals as they interact with their environment, the objectives of a learner-centred education are statements of the experiences the student should have. This view brings learner-centred educators into conflict with those educators and administrators who believe that the objectives of an education are its measurable outcomes which is the case with ABET and other systems where administrators and politicians require measures of efficiency.

Many engineering educators are influenced by the constructivist approach. At the same time, Matthews has pointed out that the constructivist approach to teaching is not unique. Many educators actively engage students in learning and do not require a particular epistemology to support their endeavours; and some would follow the steps described by Driver and Oldham [43] or the similar inquiry-based learning described by Dewey. However, the point is not to be critical of theory but to acknowledge that on the basis of theory, good practice in teaching has been developed. There is no point in arguing that teachers should have a defensible theory of learning if it is to be judged by theory, and not by the practical outcomes it causes. Moreover, it is not an excuse for discontinuing the debate or examining our operational philosophies of learning with a view to improving them.

(4) **The social reconstruction ideology** takes the view that, since society is doomed because its institutions are incapable of solving the social problems with which it is faced, education is concerned with reconstructing society. Philosophically this ideology has its foundations in John Dewey's *Reconstruction in Philosophy* and *Democracy and Education* [44, 45]. According to Schiro the social reconstruction ideology was brought to life through a split in the Progressive Education Association [46]. As might be expected it took a social constructivist view of knowledge in which knowledge is relative. The purpose of teaching is to stimulate students to reconstruct themselves so that they can help reconstruct society. Some authors see teaching as a subversive activity [47].

The principle methods of teaching are the "discussion" and "experience" group methods. In the "discussion" method the teacher elicits "from the students meanings that they have already stored up so that they may subject those meanings to a testing and verifying, re-ordering reclassifying, modifying and extending process" (Postman and Weingartner cited by Schiro [48]). In this way, a transformation of and reconstruction of knowledge occurs in response to the group process. The experience method places "the students in an environment where they encounter a social crisis and learn from those who usually function in that environment" [49]. The teacher in this technique becomes colleague and friend.

Schiro writes "human experience, education truth and knowledge are socially defined. Human experience is believed to be fundamentally shaped by cultural factors; meaning in people's lives is defined in terms of their relationship to society. Education is viewed as a function of the society that supports it and is defined in the context of a particular culture. Truth and knowledge are defined by cultural assumptions: they are idiosyncratic to each society and testable according to criteria based in social consensus rather than empiricism or logic" [50].

While the view of those who hold to this ideology may seem way outside the scope of engineering education, is it? Clearly the answer is yes. For example, some engineering educators have promoted the cause of peace engineering [51], and others social justice [52]. Langdon Winner argues that some the grand ventures that engineers engage in have anti-democratic implications which need to be thought about before they undertake them [53]. He argues that ethics education must prepare students for the political tasks they will undoubtedly face as professionals [54].

It is unlikely that many engineering educators in the western world would disagree with the view that today the primary purpose of engineering is to improve the lot of individuals and the society in which they live. Central to the achievement of that goal is engineering design, which as Bucciarelli [55] shows, is a social activity. It reconstructs society, apparently with little attention to the consequences [56]. Social reconstruction educators take the view that while "man is shaped by society and man can shape society [...] Individuals must first reconstruct themselves before they can reconstruct society" [57].

The implications of this ideology for the engineering curriculum and its teaching are profound. Without a reconstruction of teaching this would not be possible. So far that has proved impossible to achieve. If teaching is considered to be a professional activity, why is it that the activity of teaching in higher education, and all that that entails, is not considered to be a professional activity? Given that the mind is the most delicate and precious instrument that we possess, why are that many educators allowed to charge into the educational process with little more than an implicit view of learning and teaching? Could it be that they do more harm than good? Could it be that apparently good intentions lead to suicide?

1.7   DISCUSSION

Four ideologies that broadly categorize teachers are the beliefs they have about the purposes of education, the nature of knowledge, how students learn, and how the curriculum was described. They apply equally to engineering educators. These philosophies are the drivers of the teaching and learning strategies adopted. They account for some of the tensions that exist within the engineering education community, and when deeply held they are powerful resistors of change. Change is only possible when a "deep" understanding of these different philosophies is held by all the participants. In that circumstance a rational debate is possible, and the merits or otherwise of a proposed change can be evaluated. It is asserted here that many of the debates about engineering education conducted at a "surface" level which prevents understanding of different points of view, and causes any "in-depth" discussion of the aims of engineering education to be neglected. To enable "in-depth" discussion to take place it is essential that the educator has defensible theories of learning and philosophy.

There are problems with each of the ideologies. It is also doubtful if any one of the ideologies can be sustained on its own. For example, the scholar academic ideology is not concerned with learning. Curriculum concerns other than those with the discipline do not contribute to the essence of the curriculum. The role of the teacher is that of a transmitter and mediator of the knowledge contained in the discipline which the student remembers, and uses to perform mental operations. It is difficult to see how this position can be sustained in light of present-day understanding of the factors that influence learning. Furst, one of the authors of *The Taxonomy of Educational Objectives*, argues that every teacher should have a defensible theory of learning [58]. A view from which it is difficult not to assent.

Williams [59] whose analysis of the shortcomings of engineering education was little debated argued that the fragmentation of engineering into a number of specializations had deprived the curriculum of anything that was specifically engineering. Is there something that is specifically an engineering curriculum? Questions of this nature cannot be answered without an understanding of the philosophical issues involved [60]. A key question is how is a discipline formed. If the engineering discipline is simply the application of science to practical problems, should it take note of Trevelyan's argument that the exercise is pointless unless the practice of engineers is taken into account? If it is, then notice has to be taken of the affective domain, and that is prohibited by the scholar academic ideology and ignored by those who follow the social efficiency ideology, even though the authors of the cognitive *Taxonomy* also wrote a taxonomy for the affective domain [61]. Moreover, this position cannot be sustained for there is substantial evidence of the importance of the affective domain in the engineering literature [62, 63]. Both the learner-centred and social reconstruction ideologies embrace the affective domain in their attention to the whole person.

Schiro's description of the social efficiency ideology does not mention specifically the idea of competency although it may be inferred. The use of the term "competency" by engineering educators seems to go in phases. Currently, some engineering educators use it to describe outcomes, but little note seems have been taken of two views of competency that have been explored in detail in medical education [64]. The first asserts that the competency is within the person and may, therefore, be taught. That is consistent with the scholar academic ideology. The opposite view is that engineering competency is context dependent [65, 66] which is consistent with the social reconstruction ideology.

If engineering is about improving the world in which we live, then engineering educators can hardly avoid the premises of the social reconstruction ideology. There are, for example, many illustrations of courses that involve students in solving engineering problems for developing nations (e.g., [67]). It may also be argued that engineering designers are necessary members of this category, but so they are are of other categories. It is difficult to sustain the view that a professional engineering educator can be a member of one category alone. Jerry Gravander responding to this point wrote (personal communication): "every actual program of engineering education has content, intended outcomes beyond the classroom, a pedagogy and a conceived social purpose. Consequently, debates about the comparative advantages and disadvantage among various programs of engineering education are essentially debates about the proper balance among these four ideologies."

Gravander's point is illustrated by the 2015 and 2016 debates on the proposed revisions of the ABET criteria at ASEE's annual conferences. A first reaction might have been to have perceived it as a conflict between those who advocated a more liberal education for engineers and those who did not. However, it would seem more profitable to view them as demonstration of a tension between three ideologies but more particularly between the social efficiency and social reconstruction models watched by many who belonged to the scholar-academic group, none of the models being made explicit. This meant that these debates were conducted at a "surface" level when they need to be conducted at a "in-depth" level.

But as Sinclair and Tilston wrote 40 years ago, we fail to evaluate what we are doing properly because the debates lacked a proper philosophical basis [68]. Forty years on it has never been achieved for as Gravander wrote "this debate can occur only when engineering educators make their particular balances of these ideologies explicit and engage in 'deep' discussion about them. Such activity is philosophy, and the result will be an articulated and defensible philosophy of engineering education." Such is the need for professional training in pedagogy in which educators are given the opportunity to explore a variety of philosophies and learning approaches.

But finding out who we are as engineers is also inadequate for the task of the engineering educator. If teaching is considered to be a professional activity, why is it that the activity of teaching in higher education, and all that that entails, is not considered to be a professional activity? Given that the mind is the most delicate and precious instrument that we possess, why is that many educators are allowed to charge into the educational process with little more than an implicit view of learning and teaching? Could it be that they do more harm than good? Could it be that apparently good intentions lead to suicide?

1.8   NOTES AND REFERENCES

[1] Heywood, J. (2005). *Engineering Education. Research and Development in Curriculum and Instruction*, pages 55–57, Hoboken, NJ, IEEE Press/Wiley. 2, 12

[2] Macmurray, J. (1961). *Persons in Relations*. London, Faber and Faber. 2

[3] *Loc.cit.* [1, page 4–5]. 2

[4] Astin A. W. (1997). *What Matters most in College. Four Critical Years Revisited*. San Francisco, CA, Jossey-Bass. 2

[5] Pascarella, E. T. and Terenzini, P. T. (2005). *How College Affects Students Vol 2. A Third Decade of Research*. San Francisco, CA, Jossey-Bass. 2, 13

[6] Chambliss, D. F. and Takacs, C. G. (2014). *How College Works*. Cambridge, MA, Harvard University Press. 2

[7] Heywood, J., Mina, M., and Frezza, S. T. (2016). Book review. *IEEE Transactions on Education*, 59(2):154–1158. 2

[8] *Loc.cit.* [1, Chapter 7], Curriculum change and changing the curriculum. 2

[9] Sinclair, G. and Tilston, W. (1979). Improved goals for engineering education. *ASEE/IEEE Proc. Frontiers in Education Conference*, pages 252–258. 3, 16

[10] Matthews, M. R. (1994). *Science Teaching. The Role of the History and Philosophy of Science*. London, Routledge. 3, 15

[11] Matthews, M. R. (2000). *Time for Science Education. How Teaching the History and Philosophy of the Pendulum can Contribute to Science Literacy*. New York, Kluwer Academic. 3

[12] *Ibid.* 3

[13] Davis, M. (1998). *Thinking like an Engineer. Studies in the Ethics of a Profession*. New York, Oxford University Press. 3

[14] Vincenti, W. G. (1990). *What Engineers Know and How They Know It. Analytical Studies from Aeronautical History*. Baltimore, MD. The Johns Hopkins University Press. 3, 13, 15

[15] Bowen, W. R. (2009). *Engineering Ethics. Outline of an Aspirational Approach*. London, Springer-Verlag. 3

[16] Vardy, P. and Grosch, P. (1994). *The Puzzle of Ethics*, 1st ed., p. 17, London, Font/Harper Collins. 3

[17] Yokomoto, C. F. and Bostwick, W. D. (1999). Modelling: The process of writing measurable outcomes for Ec 2000. *ASEE/IEEE Proc. Frontiers in Education Conference*, 2B-1:18–22. 4

[18] Bloom, B. et al. (Eds.) (1956). *The Taxonomy of Educational Objectives. Handbook 1. The Cognitive Domain*. New York, David MacKay. 4

[19] Schiro, M. (2013). *Curriculum Theory. Conflicting Visions and Enduring Concerns*, 2nd ed., p. 4, Los Angeles, CA, Sage. 5, 7

[20] Eggleston, J. (1977). *The Sociology of the School Curriculum*. London, Routledge and Kegan Paul. 5

[21] *Loc. cit.* Ref. [14]. 5

[22] Bruner, J. (1966). *Toward a Theory of Instruction*. Cambridge, MA, Harvard University Press. 5

[23] *Loc. cit.* Ref. [14]. 6

[24] *Ibid.* 6

[25] *Ibid.* 6

[26] Tyler, R. W. (1949). *Basic Principles of Curriculum and Instruction*. Chicago, IL, Chicago University Press. 6, 14

[27] Gessner, F. B. (1975). Systems of self-paced learning. *Engineering Education*, 64(4):368. 6

[28] Keller, F. (1968). Goodbye teacher. *Journal of Applied Behavioural Research*, 1:78. 6

[29] Koen, B. V. (1971). Self-paced instruction in engineering. A case study. *IEEE Transactions on Education*, 14(1):13–20. 6

[30] *Loc. cit.* Ref. [14, p. 110]. 7

[31] Cowan, J. (2006). *One Becoming an Innovative University Teacher. Reflection in Action*, 2nd ed., Buckingham, SRHE/Open University Press. 7

[32] *Loc. cit.* Ref. [5, pages 107–110]. See also Heywood, J. (2005). *Engineering Education. Research and Development in Curriculum and Instruction*, pages 57–63, Hoboken, NJ, Wiley/IEEE Press. 7

[33] Miller, R. L. and Olds, B. M. (1994). Encouraging critical thinking in an interactive chemical engineering laboratory environment. *Proc. ASEE Annual Conference*, pages 506–510. 7

Miller, R. M and Olds, B. M. (2001). Performance assessment of Ec 2000 outcomes in the unit operations laboratory. *Proc. Annual Conference of the American Society for Engineering Education*. Session 351.3.

[34] Carberry, A., Johnson, N., and Henderson, M. (2014). A practice-then-apply approach to engineering design education. *ASEE/IEEE Proc. Frontiers in Education Conference*, pages 1845–1848. 7

[35] Woods, D. R. et al. (1997). Developing problem solving skills: The McMaster problem solving program. *Journal of Engineering Education*, 86(2):75–91. 7

[36] Boomer, G. (1992). Negotiating the curriculum in Boomer, G. et al. (Eds.). *Negotiating the Curriculum. Educating for the 21st Century*. London, Falmer Press. 7

[37] Independent study has several meaning mostly to do with study at home. In this case it refers to university degree programs in which the student negotiates what they wish to study for a degree with the university and its tutors. Such degree programs were offered at the University of Lancaster and the North East London Polytechnic in the UK during the 1980s. 7

[38] Heywood, J. (2014). The evolution of a criterion referenced system of grading for engineering science coursework. *Proc. Frontiers in Education Conference, (IEEE)*, pages 1514–1519. 8

[39] Borglund, D. (2017). A case study of peer learning in higher aeronautical education. *European Journal of Engineering Education*, 32(1):35–42. 8

[40] *Loc. cit.* Ref. [26]. 8

[41] Davies, J. W. and Rutherford, U. (2012). Learning from fellow engineering students who have current professional experience. *European Journal of Engineering Education*, 37(4):354–365. 8

[42] Wigal, C. M. (2007). The use of peer evaluations to measure student performance and critical thinking ability. *ASEE/IEEE Proc. Frontiers in Education Conference*, S3B:7–12.

[43] Driver, R. and Oldham, V. (1986). A constructivist approach to curriculum development in science. *Studies in Science Education*, 13:105–122. 8

[44] Dewey, J. (1948). *Reconstruction in Philosophy*. Boston, MA, Beacon Press. 8

[45] Dewey, J. (1916). *Democracy and Education*. New York, Macmillan. 8

[46] *Loc. cit.* Ref. [14, p. 174]. 8

[47] Postman, N. and Weingartner, C. (1969). *Teaching as a Subversive Activity*. New York, Delacorte Press. 8

[48] *Loc. cit.* Ref. [14]. 9

[49] *Ibid.* p. 186. 9

[50] *Ibid.* p. 161. 9

[51] Vesilind, P. A. (Ed.) (2005). *Peace Engineering. When Personal Values and Engineering Careers Converge*. Woodsville, NH, Lakeshore Press. 9

[52] Riley, D. (2008). *Engineering and Social Justice*. San Rafael, CA. Morgan & Claypool. www.morganclaypool.com 9

[53] Winner, L. (1986). *The Whale and the Reactor*, Chicago, IL, The University of Chicago Press. 9

[54] Cited by Durbin (2010). Multiple facets of philosophy and engineering in I. van de Poel and D. E. Goldberg (Eds.). *Philosophy and Engineering an Emerging Agenda*. Dordrecht, Springer. 9

[55] Bucciarelli, L. L. (2003). *Engineering Philosophy*. Delft, Delft University Press. 9

[56] Harari, Y. N. (2016). *Homo Deus: A Brief History of Tomorrow*. Harvill Secker. 9

[57] *Loc. cit.* Ref. [10, p. 163]. 9

[58] Furst, E. J. (1958). *The Construction of Evaluation Instruments*. New York, David Mackay. 10

[59] Williams, R. (2003). Education for the profession formerly known as engineering. *The Chronicle of Higher Education*, January 24. 10

[60] Heywood, J. (2010). Brief Encounter. A reflection on Williams's proposals for an engineering curriculum. *Proc. Annual Conference of the American Society for Engineering Education*, Paper 1296. 10

[61] Krathwohl, D., Bloom, B. S., and Masia, B. B. (1964). *The Taxonomy of Educational Objectives. The Classification of Educational Goals. Handbook II the Affective Domain*. Boston, MA, Allyn and Bacon. 10

[62] Lynch, D. R., Russell, J. S., Evans, J. C., and Sutterer, K. G. (2009). Beyond the cognitive. The affective domain, values and the achievement of the vision. *Journal of Professional Issues in Engineering Education and Practice (ASCE)*, 135(1):47–56. 10

[63] Heywood, J. (2017). *The Human Side of Engineering*. San Rafael, CA. Morgan & Claypool. www.morganclaypool.com 10

[64] Griffin, C. (2012). A longitudinal study of portfolio assessment to assess competence of undergraduate student nurses. Doctoral dissertation, Dublin, University of Dublin. 11

[65] Blandin, B. (2011). The competence of an engineer and how it is built through an apprenticeship program: A tentative model. *International Journal of Engineering Education*, 28(1):57–71. 11

[66] Sandberg, J. (2000). Understanding human competence at work. *Academy of Management Journal*, 43(3):9–25. 11

[67] Mitcham, C. and Muñoz, D. R. (2009). The humanitarian context in S. Christensen et al. (Eds.). *Engineering in Context*. Aarhus, Academica. 11

[68] *Loc. cit.* Ref. [9]. 11

CHAPTER 2

Engineering and Philosophy

William Grimson

2.1 ABSTRACT

Philosophies of fundamental and major subjects such as mathematics and science are well established and are still evolving. The philosophy of engineering in comparison, however, is at an early stage of development and has attracted little attention from philosophers with a few exceptions. This chapter considers the challenge of addressing the formulation of a philosophy of engineering by considering through a classical prism that splits the subject into five divisions, namely epistemology, metaphysics, logic, ethics, and aesthetics. Additionally, "behaviour" is used to characterize the practice of engineering.

2.2 INTRODUCTION

It should not come as a surprise to anyone that philosophy is relevant to engineering. This is simply because of the universality of philosophy as it is its nature that no domain is excluded from consideration. What might have escaped notice however, bearing in mind the preponderance of literature on the philosophy of science, is that, engineering is more in need of a philosophical examination and understanding than virtually any other human activity. The reason for this is due to the profound impact often initially unnoticed that engineering has made and continues to make on our world. This is not the place to attempt to balance the good that engineering has brought about against what might be considered the bad or undesirable. The point is that engineering has created a world that could not have been imagined by our ancestors. The impact and all the actions that contribute to that impact deserve our attention and deepest understanding.

Two hundred years ago, in 1818, Civil Engineers (UK) in drafting a document as part of an application to be awarded a Royal Chartership wrote:

> "An Engineer is a mediator between the Philosopher and the working Mechanic; and like an interpreter between two foreigners must understand the language of both. The Philosopher searches into Nature and discovers her laws, and promulgates the principles and adapts them to our circumstances. The working Mechanic, governed by the superintendence of the Engineer, brings his ideas into reality. Hence, the absolute necessity of possessing both practical and theoretical knowledge."

In this definition the implied subject Natural Philosophy would in today's terms be called Science but it is stressed that empirical science was itself an outcome of philosophy in the general sense. Also for the term Mechanic we could today translate this as technician, technician engineer, or perhaps craftsman. The point being made in the definition is that a professional engineer, in time to be called a Chartered Engineer in the UK, was required to have wide consideration of the overall context before supervising the creation of the subject of his or her largely intellectual work. It is attractive to the modern person that the "context" should not be restricted to science and empirical knowledge but rather have a broader reach. Philosophy then, in a wider sense than Natural Science, fits in well with a modern interpretation of the above definition.

As well as mediating between "Nature and the Mechanic" engineers must interact with society which requires a different form of mediation or perhaps put more accurately as negotiation. Samuel Florman's book *The Existential Pleasures of Engineering* pithily describes the challenge in two sentences noting that "professionals have an obligation to lead, but they also have a duty to serve." And "having been served, society then has no right to blame the professionals for its own shortsightedness" [1]. Often as not the arena in which this tension is resolved, or not as is sometimes the case, is the Planning and Planning Appeals stage of delivering major projects.

A question sometimes posed is: Why has engineering attracted less attention from philosophers than science? For some, the question does not arise. For example, *The Oxford Companion to Philosophy* contains a map of philosophy in which neither engineering nor technology appear [2]. And, it must be admitted that it is not easy to see where engineering or technology would be positioned. Perhaps an outer circle is required in which architecture, engineering, medicine, and other activities are included; see Exhibit 2.1. One observation that can be made is that the outer activities have a relationship with the inner ones, so medicine, for example, has relationships with "science," "law," "mind," and "social." It is plausible then that a philosophy of medicine would have some connection with the, philosophy of law, philosophy of mind, and social philosophy. Likewise, a philosophy of engineering would be expected to have something in common with the philosophy of science and the philosophy of mathematics. In this sense the philosophies of engineering and medicine can be considered complex. That doesn't imply any extra depth or a more fundamental nature of such philosophies; rather, it is just by virtue of their multidimensional composition that their respective philosophies would be expected to be complex. Another aspect is that the outer entries have a greater degree of subjectivity in how the associated activities are practised. The solid objectivity of mathematics and science is diffused into relative subjectivity through the act of creation that is at the heart of the art and craft of engineering. On another level, Popper's objective falsifiability principle has no ready counterpart in engineering—the nearest being perhaps "failure" which with its modes of failure, degrees of failure, definitions of failure, and contexts in which failure might occur are all subjective tests. We shouldn't demand too much additional insight from the circle diagram and competing options include the use of Venn diagrams and hierarchical tree structures. Suffice to say there is a desire to provide a structure showing how the multiple philosophies are related. One last point—the

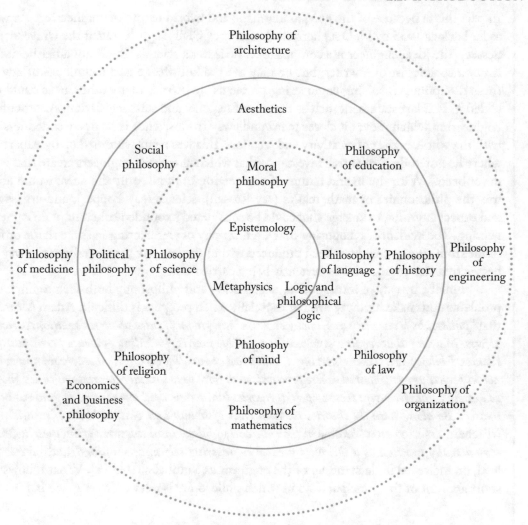

Exhibit 2.1: An adaptation of a map of philosopy by Ted Honderich in the Appendix (Map of Philosophy) p. 273 of the *Oxford Companion of Philosophy* (Oxford University Presss) to show how a philosophy of engineering might be included.

positioning of the philosophy of education is problematic as it has a different type of relationship to all the other activities than say mathematics has with science. Either each activity has its own philosophy of education (for example the philosophy of engineering education) or, alternatively, there is a global philosophy covering all of education regardless of the subject matter.

Nevertheless, the philosophy of engineering is a developing field though it is evident that it constitutes a difficult challenge as evidenced by the relatively scare literature on the subject.

In part this is because of the polyparadigmatic and hybrid nature of engineering. As was noted as far back as 1955 by N. Dougherty, a Professor of Civil Engineering at the University of Tennessee, "the ideal engineer is a composite … he is not a scientist, he is not a mathematician, he is not a sociologist or a writer; but he may use the knowledge and techniques of any or all of these disciplines in solving engineering problems" [3]. And, many other areas could easily be added to that last statement such as the role of regional and state legislation. Another feature of engineering which moves it closer to art and away from science is its open-endedness together with the whole matter of creativity and creation. This was neatly summed up by Albert Einstein where he noted that "scientists investigate that which already is; engineers create that which has never been." With the hybrid nature of engineering in mind, only the naïve would attempt to mix the philosophies of mathematics (say Russell), science (say Popper), and art (say Danto) and expect anything to emerge that could be considered a rounded coherent philosophy of engineering. The evidence is mounting that a philosophy of engineering in some shape or form will result from studies that have been produced within the last few years. Perhaps the way to put it best is that the groundwork is currently being carried out.

Finally, by way of introduction, engineering and philosophy both seek to unravel knotty problems and make headway even or especially when progress is difficult. Adam Morton stated that "*philosophy is one discipline among others, aiming to find truths about the relations between … its objects, in a way that requires evidence from fallible sources, including evidence pre-digested by other sciences. Philosophy is like engineering … concerned above all with topics where theory and evidence are not in perfect agreement, and where practical needs force us to consider theories which we know cannot be exactly right. We accept these imperfect theories because we need some beliefs to guide us in practical matters. So along with the theories we need rules of thumb and various kinds of models*" [4]. Carl Mitcham asserted that "*because of the inherently philosophical character of engineering, philosophy may actually function as a means to greater engineering self-understanding*" and taking this as a lead, an increased understanding of the engineer as a global citizen [5]. What follows next is a short account of the principal ways in which philosophy is relevant to engineering.

2.3   THE FIVE MAIN BRANCHES OF PHILOSOPHY (CLASSICAL)

As a basic starting position and not relying on any emerging philosophy of engineering, it is worth recalling Ludwig Wittgenstein's view that "*Philosophy is not a theory but an activity.*" In engineering, the question then posed is what areas of activity might a philosophy of engineering emerge or be articulated. It is attractive to consider the following five branches that have been thought and written about for centuries to put some structure on such philosophical activity, namely: Epistemology, Metaphysics, Ethics, Logic, and Aesthetics. These five branches are summarised in Table 2.1 and expanded upon in the following sub-sections.

Table 2.1: Summary of the five classical branches, some questions they address, and the categories within

	Description	Some Main Questions	Categories (Examples)
Epistemology	Process by which knowledge is gained	What is knowable? How is it acquired? Is it valid?	Rationalism, empiricism, logical-positivism
Metaphysics	Study of reality that is beyond the physical	Existence of God, the soul, and the afterlife. What is existence?	Investigation into the nature of reality, uncovering what is ultimately real
Ethics	Study of moral value, right, and wrong	Placing value to personal actions, decisions, and relations	Moral theory, virtue ethics, religion and ethics, applied ethics
Logic	Study of right; reasoning	Tool used to study other philosophical categories	Propositional logic and predicate calculus, quantum logic, temporal logic
Aesthetics	Study of art and beauty	What is the relationship between beauty and art? Are there objective standards? Is beauty in the eye of the beholder? Form versus function	Aesthetics in the arts, aesthetics in the sciences, aesthetics in engineering (design)

Epistemology   seeks to understand the distinction between different forms of knowledge (rational as in mathematics, empirical as in most if not all of science, etc.); to consider how knowledge is acquired, recorded, organised, encoded, maintained, transmitted and used; and to provide a platform by which the provenance and limits of the applicability of knowledge may be evaluated and understood. Tacit knowledge arises where the communication whether by written word or orally fails to tell the whole story. This arises more in the craft-derived end of engineering rather than in the more formal engineering science.

Metaphysics   considers the question of what is reality, including abstract concepts such as substance, knowing, time and space as well as relationships. Metaphysics also includes ontology, mereology (relationship of parts to whole and the relationship between parts within a whole), and teleology considerations. Ontology amongst other things addresses the nature of being and by extension therefore what it means to be an engineer.

Ethics   examines the determinants of appropriate behaviour, placing value on personal actions, decisions and relations; the impact of legislation and professional code of ethics (Hippocratic oath and equivalent ones for engineers and scientists; societal concerns); personal moral compass and concept of virtue; and cultural influences.

Logic   studies concepts of "right reasoning," forms of logic (e.g., temporal logic), role of logic in building conceptual models, the role of logic in how knowledge is deployed.

Aesthetics   examines the distinction between "values" in arts, science, and engineering: the tension or even dialogue between form and function. Since engineering involves designing and making things that did not previously exist, aesthetic issues are raised at each departure.

An appreciation of the uncertainty and ambiguity of knowledge is as important to engineers as it is to the medical profession. This is a topic explored later in Chapter 3 by Jerry W. Gravander. Dealing with ambiguity in requirements engineering is a well documented problem area especially in software engineering. Decision making based on incomplete data sets is often a challenge where opting out of making a determination might not be an option. George Bernard Shaw, admittedly with medical doctors in mind, aimed a volley at professions and accused them of conspiring against the laity. But engineers and the medical profession often have to proceed with incomplete data or uncertain conditions where to do otherwise would risk an unwelcome outcome. One way of countering Shaw's view is by fostering better communications between the professions and those they serve. With current hotly contested debates in international society on topics such as fracking, it is of paramount importance that the engineering profession acts in an open and ethical manner: trust once lost is not easily regained. It is not just the beauty or otherwise of cars, bridges and buildings that concern designers and end users. Many environmental debates have an aesthetic dimension, such as the positioning of wind turbines or tall pylons distributing electrical energy in places of natural beauty. The point being made here is that in almost every engineering endeavour you may wish to consider it is not difficult to observe the ethical and aesthetic dimensions of what is involved.

In systems engineering there is a need to examine and describe the methods and methodologies for building ontologies. These ontologies are normally formal representations of a set of concepts within a defined domain together with the relationships between those concepts. Both abstract and non-abstract concepts would typically need to be addressed. Those charged with managing large inventories, such as are found in the aerospace industry, have had to deal with ontological challenges. The philosopher Peter Simons has made a study of the relevance of ontology to engineering [6]. Engineers have clarified their thinking in dealing with functional decomposition using mereology (a treatment of how parts relate to the wholes they contribute to and shape). Teleological considerations are generally avoided in scientific and engineering work. Teleology deals with the contention that nature in some manner strives toward a particular end. As an example, James Lovelock's Gaia hypothesis is considered to be teleological. At the very least engineers looking at biological systems and the associated literature should be aware of the usage at times of teleological language.

The view presented in this chapter is that the five branches are essentially orthogonal and are therefore all necessary if a full account of a philosophy of engineering is to emerge. Table 2.2 considers some of the main epistemology theories and their relevance to engineering to demonstrate how some traction might be gained.

A more lengthy description of the main branches of philosophy set in historical perspectives would allow the development of philosophy to be seen in the context of the parallel expansion of what man has discovered or created in mathematics, science, engineering, technology, and other domains. That is well outside the scope of this short chapter, nevertheless the above descriptions should illustrate the point that engineering is inherently philosophical in that its activities do indeed impact on and depend on epistemological, metaphysical, ethical, logical, and aesthetic deliberations. A point made by the author previously is that any large project, especially a ground-breaking one such as the design and construction of the Crystal Palace masterminded by Joseph Paxton for the Grand Exhibition of 1851, can be analysed in terms of how it addressed what otherwise might be considered philosophical activities [7]. Indeed, in addressing "knotty problems" engineering can be said to be an exercise in applied philosophy. Engineering might not reach the heady intellectual heights of Hegel, Heidegger, or Habermas but it certainly strives to find solutions to some world problems.

2.4   BEHAVIOUR

From another perspective and complementary to the five classical branches of philosophy outlined above, the Oxford English Dictionary (OED) entry for philosophy gives: *In extended use: a set of opinions or ideas held by an individual or group; a theory or attitude which acts as a guiding principle for behaviour; an outlook or world view.*

Applying the OED definition to engineering means characterizing in some manner (a) what opinions and ideas are held by engineers and how they originated, (b) what attitudes prevail in determining behaviour and how they arose, and (c) the general outlook that is commonly held amongst engineers, what shapes it and how it reflects, or not as the case might be, societies views. How opinions, attitudes, and outlooks are formed in the mind of an engineer as they develop their career depends on their upbringing, education, early work experience, mentors, and society in which they work. Many of the chapters in *The Engineering Business Nexus* address this question [8]. Nan Wang and Bocong Li give an account of engineers in China, Eddie Conlon describes the engineer as a captive of capitalism, and Russell Korte examines the nature of workplace culture into which new graduates socialize. There are many other chapters in this book which taken together paint a very complex picture of what makes an engineer the person and member of society that they are and how they behave. Another dimension is uncovered in Christelle Didier's chapter in *Engineering in Context* in which religious and political values can form an engineering ethos and determine behaviour and outlook [9].

Table 2.2: Brief statement of the engineering dimension to various theories of epistemology (*Continues.*)

Epistemology Theory	Definition*	Engineering Dimension
Empiricism	Based on experience, a result of observation. The doctrine which regards experience as the only source of knowledge.	Very much to the fore in engineering disciplines.
Constructivism	A sub-set of empiricism and concerned with how we learn.	Relevant all stages of the formation of an engineer.
Behaviorism	Whilst often defined in psychological terms it is basically concerned with conditioning.	Relevant to how engineers work in groups or organizations.
Rationalism	Ideas not derived from our experience or observation. Based on pure thought. A theory (opposed to *empiricism* or *sensationalism*) which regards reason, rather than sense, as the foundation of certainty in knowledge.	Clearly some knowledge is rationalist in nature but for the engineer subsequent justification from experience is valued. Mathematics is a good example, and is of direct relevance to engineers.
Positivism	The only authentic knowledge is scientific knowledge. Or more generally, any of various philosophical systems or views based on an empiricist view of science, particularly those associated with the belief that every cognitively meaningful proposition can be scientifically verified or falsified.	Engineering could never have developed based on such a narrow definition of knowledge. Planes flew before engineers had available sound aerodynamic scientific "knowledge." Failure rather than falsifiable is the key engineering concept here.
Logical positivism	Also called logical empiricism, rational empiricism, and includes the verifiable principle; its alternative (anti-logical positivism) is Popper's falsifiability principle. Logical positivism, the name given to the theories and doctrines of philosophers active in Vienna in the early 1930s (the Vienna Circle), which were aimed at evolving in the language of philosophy formal methods for the verification of empirical questions similar to those of the mathematical sciences, and which therefore eliminated metaphysical and other more speculative questions as being logically ill founded.	Engineers can work satisfactorily without considering this theory?

Table 2.2: (*Continued.*) Brief statement of the engineering dimension to various theories of epistemology

Epistemology Theory	Definition*	Engineering Dimension
Idealism	What we perceive as the external world is in some way an artifice of the mind.	Not held to be relevant by most engineers it is conjectured!
Existentialism	Existentialism considers that action, freedom and decision as fundamental to human existence. Underlying themes and characteristics, such as anxiety, dread, freedom. To a large extent, existentialism is at odds with the western rationalist principles: it takes into account human beings' actions and interpretations however irrational they may seem.	Increasingly important perspective for engineering to take into account the human dimension to a greater extent than at present
Philosophy of science	Hypothesis, prediction, followed by experimentation and supporting or denying the hypothesis.	Engineering both contributes to knowledge thus gained and inherits knowledge directly from the work of scientists.
Philosophy of engineering	The opinions about or ideas as to what constitutes knowledge relevant to engineering which in turn acts as a guiding principle for behavior.	
Transcendental idealism	Unlike idealism, does not claim that the objects of our experiences would be in any sense *only* within our minds. Perception is *influenced* by the categories and the forms of sensation, space, and time, which we use to understand the object.	This is relevant, surely, to what is happening at design stages where society and end-users must be considered together with many other constraints being part of the context.

* Based on definitions in http://en.wikipedia.org/wiki/Epistemology.

The matter of opinions, behaviour and outlook is inevitably linked to the five branches of philosophy (classical) discussed briefly above. In particular epistemology and ethics, following the discussion by Vardy and Grosch (Chapter 17), can be viewed in terms of what "is" and what "ought" [10]. As already pointed out "scientists investigate that which already is; engineers create that which has never been." The behaviour of scientists is concerned primarily with adding to our store of knowledge of what exists. Whereas the engineer, using the knowledge so gained, embarks on creating something new, and where implicitly a decision has been made that the something "ought" to be. If in the past the decision making by engineers about what ought to be created carried little in the way of ethical consideration the same cannot be said today. Environmental and climate change as just two issues permeate the discourse amongst and between society and the engineering community in a wide range of developments. The switch from fossil fuels to sustainable sources of energy, the schemes to reduce waste (particularly plastic), the efforts to build low-energy housing, and the need for more efficient and effective transport system are all concerned with what "ought" to be. Using the earlier notion of a professional engineer as a mediator we can add the role of mediation between society and what is achievable where the "ought" is now a negotiated position. The Aarhus Convention based on the three principles: Access to Information; Public Participation in Decision-making; and Access to Justice in Environmental Matters, and in general planning laws in each jurisdiction are just some of the ways in which this negotiation takes place [11]. Lastly, the engineering bodies responsible for the education of engineers recognise the role of ethics in the formation and subsequent behaviour of members of the profession as evidenced by the criteria in the accreditation of degree programmes. And there has been a call from some quarters for laypersons to be included on accreditation panels.

One question that requires a response is "if engineering is inherently philosophical why would it be of any benefit for an engineer to study philosophy." There are a number of lines of response possible. First and foremost "critical thinking" is considered to be a highly desired attribute of a graduate. Part of that attribute is the skill or habit of thinking outside the domain normally inhabited by the graduate. Likewise, as Michael Brooks wrote in the *New Scientist* there is a recognition that "we need agile thinkers rather than just more science, technology, engineering, and maths graduates" [12]. In terms of undergraduate engineering education the engineering educationalist David Goldberg has written about the broken curriculum and has identified the need for the inclusion of qualitative thinking which he states has its roots in philosophy. What better toolkit than that made available by philosophy to aid the "thinking engineer;" and one that is generic or universal and one that has been sharpened by use across many fields. In summary, the British philosopher Jonathan Rée put it very succinctly when stating that *"philosophy is about learning to be aware of problems in your own thinking where you might not have suspected them"* [13].

2.5   CONCLUSION

The lens of philosophy can help engineers see things in a more complete manner; however, there are other aspects that should positively concern us and fall under the umbrella of education. As engineers are formed first through education and second through experience, it is clear that both general philosophies of education and philosophies of engineering education are relevant. There is another reason why this topic is of current interest namely the curriculum problem by which educational programmes are under immense pressure to deliver across so many components as per accreditation requirements. In short, accreditation criteria look to (a) knowledge and understanding (mathematics, sciences, and technologies); engineering analysis (including the ability to identify and formulate problems); engineering design; investigations (including the ability to design experiments); understanding the need for high ethical standards in the practice of engineering together with responsibilities of the profession toward people and the environment; working in multi-disciplinary settings; interaction with society at large. It is evident that the breadth and depth of the educational experience to reach what is required at bachelor or master's level represents a huge challenge for both the student and engineering school providing the education. Pedagogical initiatives underpinned by a coherent philosophy of (engineering) education must surely be the concern of all educators.

To some extent engineering has always searched for an identity, unlike the case of the practitioners of other professions (law, medicine, architecture). The problem is not helped by the loose way in which the word "engineer" is used (from the unqualified to the qualified at various levels). To that end, some recent writings that explore how opinions and ideas are formed, attitudes that guide behaviour, and in general how outlook is shaped, are or should be of interest to professional engineers. Arising from a number of workshops held in the UK and the U.S., Diane Michelfelder, Natasha McCarthy, and David Goldberg published *Philosophy and Engineering: Reflections on Practice, Principles and Process* [14]. The volume explores the ontological and epistemological dimensions of engineering and "exposes the falsity of the commonly held belief that the field is simply the application of science knowledge to problem solving." Amongst other topics discussed in this volume, ethical considerations are prominent including the obligations of professional capabilities, and the role for an engineering Hippocratic oath. Bearing in mind that culture plays a role, American, Chinese, and European perspectives are presented in *Engineering, Development and Philosophy* [15]. Together with other material (see Bibliography by Heywood et al. [16]) these books illustrate the richness of what often is unobserved in engineering or is just plainly taken for granted. It is not just the non-engineer who doesn't observe. In the first instance, it is the responsibility of the engineer and the profession in general to understand themselves. Philosophy can help in that last regard.

To conclude, a philosophy or even philosophies of engineering can be articulated by describing the "behaviour" of engineers through the perspectives of epistemology, metaphysics, ethics, logic and aesthetics. And in the spirit of this volume, the philosophy or philosophies of engineering can then inform the philosophy of engineering education all the better both to reinforce that which is deemed good and address those areas considered poor. Who decides such matters seems a reasonable question with society a safe answer. But first it is the engineers themselves who need to respond with the benefit of increased self-knowledge as Carl Mitcham has asserted [17].

2.6   NOTES AND REFERENCES

[1] Florman, S. (1976, 1994). *The Existential Pleasures of Engineering.* New York, St. Martins Press. 18

[2] Honderich, T. (Ed.) (2005). *The Oxford Companion to Philosophy*. Oxford University Press. 18

[3] Dougherty, N. Quoted in https://engineering.purdue.edu/Engr/AboutUs/News/publications/EngineeringImpact/2010_2/COE, May 2019. 20

[4] Morton, A. (2001). Philosophy as Engineering. Chapter 3 in Bo Mou (Ed.), *Two Roads to Wisdom?—Chinese and Analytic Philosophical Traditions*. Open Court Publishing Company. 20

[5] Mitcham, C. (1998). The importance of philosophy to engineering. *Technos*. XVII/3. http://campusoei.org/salactsi/teorema02.htm 20

[6] Simons, P. (2013). Varieties of parthood. Ontology learns from engineering in Michelfelder, D., McCarthy, N., and Goldberg, D. E. (Eds.), *Philosophy and Engineering: Reflections on Practice, Principles, and Process*. Dordrecht, Springer. 22

[7] Grimson, W. (2007). The philosophical nature of engineering—a characterisation of engineering using the language and activities of philosophy. *Proc. Annual Conference American Society for Engineering Education*. CD Paper AC2007–1611, (14 pages). 23

[8] Christensen, S., Delahouse, B., Didier, C., Meganck, M., and Murphy, M. (Eds.) (2019). *The Engineering-Business Nexus*. Springer. 23

[9] Didier, C. (2009). Religious and political values and the engineering ethos in Christensen, S., Delahouse, B., and Meganck, M. (Eds.), *Engineering in Context*, Chapter 3, Aarhus, Academica. 23

[10] Vardy, P. and Grosch, P. (1994). *The Puzzle of Ethics*, Chapter 17, London, Harper Collins. 25

[11] Aarhus Convention. http://ec.europa.eu/environment/aarhus/index.htm, May 2019. 26

[12] Brooks, M. (2013), Invest in minds not maths to boost the economy. *New Scientist*, Issue 2948, Reed Business Information Ltd. 26

[13] Rée, J. in essay titled *In Defence of Heidegger*. https://www.prospectmagazine.co.uk/arts-and-books/in-defence-of-heidegger#.UyGh3lFdVRc, May 2019. 26

[14] Michelfelder, D., McCarthy, N., and Goldberg, D. (Eds.) (2013). *Philosophy and Engineering: Reflections on Practice, Principles, and Process*. Dordrecht, Springer. 27

[15] Christensen, S. H., Mitcham, C., Li, B., and An, Y. (Eds.) (2012). *Engineering, Development, and Philosophy—American, Chinese, and European Perspectives*. Dordrecht, Springer. 27

[16] Heywood, J., Carberry, A., and Grimson, W. (2011). A select and annotated bibliography of philosophy in engineering education. *Proc. ASEE/IEEE Frontiers in Education Conference*, PEEE:1–26. 27

[17] Mitcham, C. (2014). The true grand challenge for engineering: Self-knowledge. *Issues in Science and Technology*, 31(1). 27

CHAPTER 3

Philosophy of Engineering as Propaedeutic for the Philosophy of Engineering Education

Jerry W. Gravander

3.1 ABSTRACT

The philosophy of engineering has implications for the philosophy of engineering education. This chapter develops a point in the philosophy of engineering, namely, that the results of engineering practice are inescapably and inalterably uncertain. It then briefly explores what this implies for the philosophy of engineering education, particularly some ways in which engineering education should change in light of the uncertainty in engineering practice.

3.2 INTRODUCTION

I have given several papers over the years at the annual conferences of the American Society for Engineering Education (ASEE) and elsewhere about the content and structure of engineering education, [1, 2] are representative, and these have rested on two principles:

- first, engineering education should be more than vocational training; and

- second, the complexity of the physical and social environment requires a complex integration of technical and non-technical factors in engineering design.

The first of these is a philosophical observation about engineering education, and the second is an empirical conclusion about engineering practice. I certainly have not been alone in accepting these principles. As shown by Bruce Seeley's work in the history of engineering education [3], they have been at the heart of every report on the engineering curriculum commissioned by ASEE, and they are manifest in ABET's accreditation criteria for engineering.

As commonplace as they are, however, we still can ask why. Why shouldn't engineering schools be content with producing the best technicians they can? Why must engineering practice be based on integrative designs? In developing this paper, I came to believe that the philosophy of engineering [4] can provide answers.

Gregory Basset and John Krupczak, Jr., argue in Chapter 4 that a primary difference between science and engineering is their product.

- Science moves from data about a set of specific existent entities to an abstract theory that unifies these entities with all other entities in the world that share the common characteristic(s) identified by the theory. A theory can predict existent entities, but it cannot create them.

- Engineering moves from abstract considerations about intended function and the range of available means to achieve it to a particular design that communicates an executable plan for providing the intended function. Although the execution of a design utilizes existent entities, it creates entities and adds them to those already existent in the world.

Basset and Krupczak suggest that this point from the philosophy of engineering has an implication for the philosophy of engineering education, namely, that the engineering curriculum should develop students' ability to innovate and explore alternative designs for achieving a specific intended function—in short, think divergently—before converging from their abstract starting point to a particular problem solution. In other words, engineering students should learn how to look for alternative and equally functional solutions to their design problems. I would add that they also need to learn how to go about selecting one and then how to communicate it to others as an executable plan.

In the remainder of this chapter, I will argue for these same conclusions from another point in the philosophy of engineering, one that I consider central, and describe some of what this implies for the philosophy of engineering education.

3.3   THE PHILOSOPHY OF ENGINEERING IS NOT THE PHILOSOPHY OF SCIENCE

Although I think it is self-evident that engineering is not science, I will briefly argue for this point. Science is abstract, idealizing, and reductionistic, and its products are theoretical concepts. As I tell students, the Ideal Gas Law describes how gases would behave if only they were not real, but of course they are; no real gas follows this law. Indeed, my freshman chemistry book clearly stated this and included such useful empirical approximations as van der Waals equation. Similarly, inertial motion would exist only in a universe populated by a single object in a perfect vacuum, absolute zero is a useful theoretical concept but cannot be reached in practice, and so on. In contrast, engineering is concrete, pragmatic, and holistic, and its products are actions. Unlike science, which artificially restricts problems, engineering deals with the real complexity of a world in which one can never know whether or not all of the relevant factors and variables have been identified, let alone treated correctly. Science seeks truth about the world, and there can be only one. Engineering seeks optimal solutions to problems in the world, and there are always multiple, equally acceptable solutions depending on the assumptions about the unknown factors. No engineer should be surprised by any of this.

It follows that the philosophy of engineering cannot be the same as the philosophy of science [5].

3.4   THE PHILOSOPHY OF ENGINEERING NARROWLY DEFINED

For the purposes of the central argument in this essay, I will narrowly define the philosophy of engineering as a field of inquiry, and I will do so in parallel with a narrow definition of the philosophy of science.

As found since the early-20th century in the pages of philosophy of science journals, as well as numerous textbooks and monographs, the philosophy of science primarily focuses on such issues as the logic of scientific law and theory formulation and justification, the epistemology of science's theoretical constructs, and the structure and logic of scientific explanation. The philosophy of engineering as I am defining it parallels this focus by exploring the logic of engineering judgment, the epistemology of engineering designs, and the structure and logic of engineering practice.

In contrast to the philosophy of science's relatively long history as a field of inquiry, the philosophy of engineering has only recently begun to emerge as a field of inquiry. The flagship journals for the philosophy of science, *Philosophy of Science* and *The British Journal for the Philosophy of Science*, are almost 80 years old and over 50 years old, respectively, and the professional associations affiliated with these journals predate them. The philosophy of engineering as a field of inquiry dates from the mid-2000s with workshops in the United States in 2006 and the Netherlands in 2007. The proceedings of the latter being published in 2010 [6].

3.5   PHILOSOPHY OF ENGINEERING IS NOT PHILOSOPHY OF ENGINEERING EDUCATION

Despite the above differences in content between the philosophy of science and the philosophy of engineering, their general frameworks are similar in their narrow focus. The philosophy of science as a field encompasses the logical and epistemological characteristics of the scientific method and its theoretical products. Similarly, the emerging field of the philosophy of engineering encompasses the logical and epistemological characteristics of engineering method and its practical products. References to the philosophy of engineering are not frequently encountered in discussions about engineering education. Indeed, the use of the words "philosophy of engineering" in the above sense is rare at ASEE conferences; a full text search on "philosophy of engineering" in the programs and proceedings for the 1996–2013 annual conferences of ASEE yields only eight hits—one session and seven presentations [7]. In general, these papers developed points in the philosophy of engineering for the purpose of critiquing and reforming the philosophy of engineering education.

In contrast to this relatively narrow and strict definition for philosophy of engineering, I am using "philosophy of engineering education" in a broad and loose sense. All reflection on engineering education—its goals and objectives, curriculum, and pedagogy—counts as the philosophy of engineering education. In other words, this paper is not treating the philosophy of engineering education as a field *per se*, but rather is a diverse body of thought about all aspects of engineering education.

3.6   ENGINEERING UNCERTAINTY

This brings me to the point in the philosophy of engineering that is the focus of this chapter: the results of engineering practice are inescapably and inalterably uncertain.

First, engineering products are technically uncertain. We do not know until we implement them whether they will technically succeed or fail, stand or fall, fly or crash, compute correctly or not. I am not the first person to comment on this, and several reasons for this uncertainty have been noted. I will mention two.

Elting E. Morrison in his book, *From Know-How to Nowhere: The Development of American Technology* [8], argues that until the late-19th century, American engineering was always done in the absence of the knowledge needed to complete the project. His paradigm case was canal building. The engineers knew canals could be built—they existed elsewhere in the world—but there had been a generational break in the transmission of knowledge about how to do so. So, they simply started building canals with the expectation they could figure it out as they went along.

Morrison claims this changed with the establishment of the GE research and development labs in the 1890s. I am not so sure. In the late 1970s I worked on a study of the Clinch River Breeder Reactor Plant, a proposed prototype commercial liquid sodium-cooled breeder reactor. The central safety question was what to do about a possible loss of coolant accident. The project had been carrying two designs forward. One design had a "core catcher" in case there was a loss of coolant accident and subsequent core meltdown, and the other design had no provision for such an accident. I was at the technical briefing when the project director said the core catcher design was being dropped because a loss of coolant accident had been judged "less probable than improbable." Someone in the audience asked, "But surely you have some provision in case the worst possible thing that could happen happens?" The answer was, "Yes, we will vent the containment dome and filter the released gases," and although the project director admitted they had no idea how to make such filters, "the difficult we do today, and the impossible we do tomorrow." At this point, hundreds of millions of dollars had been spent, and they had no idea how to finish.

A second source of technical uncertainty is the nature—that is, the logic and epistemology—of the engineering method. In his *Definition of the Engineering Method* published by ASEE [9], Billy Vaughn Koen defines engineering method as "the strategy for caus-ing the best change in a poorly understood or uncertain situation within the available resources." The Rule of Engineering he states is to "do what you think represents best practice at the time you decide." Koen notes that the engineering method rests fundamentally on heuristics, that is, uncertain and "fuzzy" rules of thumb. Koen draws two explicit consequences from his analysis:

- multiple, equally justifiable engineering solutions, based on different sets of heuristics and different understandings of the state-of-the-art, are not only possible but likely for a given problem; and

- the path to engineering progress is a refinement and clarification of heuristics that is driven by engineering failures.

Henry Petroski makes the latter point in his *To Engineer is Human: The Role of Failure in Successful Design* [10]. The realities of the engineering method mean that engineers cannot even conceive of, let alone eliminate, all the possible ways in which a design can fail, and yet the designs will be implemented. I would add, they also cannot conceive of all the ways their designs can succeed, nor the ways in which some of these successes in a technical sense are failures in a broader sense.

This brings me to the second type of uncertainty associated with engineering—the uncer-tainty of the social and physical consequences of the products of engineering. Lest we forget, there is no such thing as pure engineering. Engineering is undertaken solely for the purpose of implementation. My friend Arthur Sachs from the Colorado School of Mines stood up in every ASEE session he attended and said, "Until you engineering educators understand that you are graduating change agents who will alter the world system in profound ways, you will never get engineering education right." Arthur had a dramatic flair, but his point is essentially correct. The products of engineering methodology change the social and physical environment in poorly understood and extremely difficult to anticipate ways.

The combination of uncertainties in both the technical designs and social-physical conse-quences of engineering led Michael Martin and Roland Schinzinger [11] to conclude that the engineering process has the same logical and epistemological structure as an experiment—and that engineering literally is a jointly technical-social experiment on human subjects. I take a somewhat more conservative stance (which is atypical for me) and hold that there are important analogies between the engineering process and experimentation on human subjects rather than an equivalency, but I reach the same primary conclusion as Martin and Schinzinger. Human participants in an experimental situation are owed the right of informed consent. In the case of engineering, who exactly will do the informing? I submit that the only plausible candidates are the engineers—if not solely, then as major players [12].

One possible counterargument to the analysis above is that the implementation of engi-neering designs could always be delayed until the underlying science was made complete enough, the design principles determinate enough, and the impact models detailed enough. However, this counterargument ignores the fact that engineering, unlike science, is always practiced under what William James labeled "forced choice" conditions [13]. James developed this concept in the context of religious belief, but it can be generalized. He points out that people typically believe that decisions have three options: positive, negative, and suspended judgment. In James' example regarding belief in God, these three options would be theism (actively believing in God), atheism (actively disbelieving in God), and agnosticism (withholding commitment to either belief or disbelief). He points out that in the case that God exists, atheism and agnosticism have exactly the same consequences, namely the loss of whatever benefits derive from believing in God. James shared the Pragmatic Theory of Meaning with the other American Pragmatists, and this theory holds that two concepts that have the same consequences are actually the same concept, no matter how different they may appear. James concludes that regarding belief in God, people have only two "live" options, namely belief (theism) and disbelief (atheism/agnosticism). The suspended judgment option is not available, and consequently the choice is "forced" between belief and disbelief. Without having to accept this example, James' concept of "forced choice" can be generalized. Take friendship as another example. Suspending your judgment that a person is your friend deprives you of this friendship just as certainly as disbelieving that he or she is your friend. Whenever conditions exist under which the consequences of suspended judgment are the same as the consequences of either the positive or the negative judgment, a person does not have suspended judgment as a "live" option. They have only the "forced choice" between yes and no.

Engineering projects are time-constrained in the sense that the financial, social, resource, market, etc., conditions surrounding them are such that if the projects are not started by a given time, they will not be started at all. There is a time, T, such that delaying the go/no-go decision past T has the same consequences as deciding no-go prior to T. In other words, at time T during a project's planning phase, there is a "forced choice" between yes and no with respect to starting the project. But what if uncertainties regarding the project's underlying scientific basis, engineering design, and/or impact still exist at time T? The engineer and his or her client or employer for the project either need to act in the face of uncertainty or abandon the project. Scientists can always suspend judgment if there are uncertainties regarding a hypothesis, but engineers cannot suspend judgment when there are uncertainties regarding a proposed project. I take this to be an essential characteristic of engineering judgment. It is not a mistake on an engineer's part to tolerate the uncertainty that follows from the "forced choice" structure of engineering judgment, but rather it is simply a reality of engineering practice.

3.7   IMPLICATIONS OF ENGINEERING UNCERTAINTY FOR THE PHILOSOPHY OF ENGINEERING EDUCATION

So, what points within the philosophy of engineering education do I infer from my excursion in the philosophy of engineering?

First, design is not just an important component of engineering education, it is the paramount component. Moreover, the capstone project "subject to realistic engineering constraints" required by ABET is not enough. In the past I have argued that the capstone project would become sufficient only if the constraints addressed by the students were multidimensional enough [14]. However, I now believe this would not be sufficient. Students should have multiple design experiences that include the implementation step that is an inevitable part of post-graduation engineering work—fly planes they have designed, turn the key on their pilot plants, build dams, etc. Only then will student projects include a real possibility of "design to failure," and students should confront this possibility before they are released on the world. This type of truly real-world experience possibly could be included in the curriculum if it were more than four academic years long [15]. However, a better approach might be to copy other professions that rest as much on practice as on theory, for example, medicine, and require an internship/apprenticeship after the completion of degree work, where this internship/apprenticeship would have to be completed satisfactorily before independent engineering practice of any kind could be undertaken.

Second, students should have opportunities to experience the phenomenon of multiple, equally justifiable solutions to a problem, and they should be helped to develop strategies and methods for resolving these "on the job." Putting this into practice would require significant changes in many, if not most, engineering courses, where there is only one right answer and negotiable partial credit.

Third, students should receive preparation for their roles in an informed consent procedure within a democratized technological decision process. This has not only a communication aspect—and this extends beyond the engineering communication students now encounter—but also an understanding of the social and political processes by which public policy and social acceptance regarding technology are shaped. Moreover, as argued by John Heywood and Michael Lyons [16], the communication aspect of a democratized technological decision process will require a common language shared by engineers and a lay public into which the engineers can translate their technical language, and this, in turn, will require educating the public for engineering and technological literacy, an activity in which engineers will have a role. Engineering education will need transformation along a number of dimensions for students to even be aware of what society needs from them as practicing engineers, let alone be prepared to play a role in delivering it.

3.8   BROADENING THE ARGUMENT

My argument above drew on William James' notion of forced choice and the Pragmatic Theory of Meaning shared by James and other American Pragmatists, including Charles S. Peirce, George Herbert Mead, and John Dewey. The thought of the American Pragmatists and later philosophers in the pragmatic tradition provides a rich set of philosophical perspectives from which to explore the nature of engineering knowledge and practice, and the results of such exploration have significant implications for the philosophy of engineering education. Volume 2 [17] of this series contains chapters engaged in such exploration. Russell Korte in Chapter 1 surveys a selection of ideas drawn from the thought of philosophical pragmatists and briefly explores their application to engineering practice. Mani Mina focuses on John Dewey in Chapter 2, the most influential American Pragmatist, and develops several Deweyan conclusions regarding engineering education. In Chapter 3, Steven T., Frezza, and David A. Nordquest focus on Bernard Lonergran's pragmatic theory of knowledge and derive several implications for engineering pedagogy. Despite their different paths, all three papers reach the conclusion that engineering design should become the predominant thread in engineering curricula, as opposed to the current predominance of analysis, and the variegated details of the arguments in the papers suggest several transformative improvements for engineering education.

3.9   CONCLUSION

My goal in this chapter was to show that the philosophy of engineering is propaedeutic for the philosophy of engineering education. The discussion above is sufficient to establish what philosophers call an existence proof. The fact that the point from the philosophy of engineering that I develop from the uncertainty of engineering has implications for the philosophy of engineering education establishes the general claim. My brief overview of the papers in Volume 2 of this series provide further support. Moreover, I have defined "philosophy of engineering education" broadly enough to encompass all areas of inquiry within ASEE. It follows that the philosophy of engineering should become a major theme within ASEE and its annual conferences, unlike what is presently the case. The philosophy of engineering provides an important framework within which to think about the philosophy of engineering education.

3.10   NOTES AND REFERENCES

[1] Gravander, J. (2004). Toward the "integrated liberal arts:" Reconceptualizing the role of the liberal arts in engineering education. *Humanities and Technology Review*, 23:1–18. 31

[2] Gravander, J., Luegenbiehl, H., and Neeley, K. (2004). Meeting ABET criterion 4: From specific example to general guidelines. *ASEE Annual Conference Proceedings*. https://peer.asee.org/meeting-abet-criterion-4-from-specific-examples-to-general-guidelines   31, 39

[3] Seeley, B. (2005). Patterns in the history of engineering education reform: A brief essay. *Educating the Engineer of 2020: Adapting Engineering Education to the New Century*, pages 114–130, National Academy of Engineering, Washington, DC, The National Academies Press. 31

[4] There are at least three ways to use the term "philosophy of engineering." The first is to refer to philosophical thought about engineering. Whenever a person reflects philosophi-

cally about engineering, the person *ipso facto* is engaging in the philosophy of engineering and the results from the reflection count as philosophy of engineering. The second is to refer to a field of philosophical inquiry regarding engineering that has a group of practitioners whose members identify themselves as working within this field. The third is to refer to a specific philosophical theory about engineering. I will not be using "philosophy of engineering" in this third way. I will distinguish between the first and second usages by using "philosophy of engineering" and "philosophy of engineering as a field," respectively. 32

[5] The philosophy of engineering also is not the philosophy of technology, nor is it subsumed within the philosophy of technology. Explicating these points in detail would take me too far from the central argument of this chapter. 33

[6] van de Poel, Ibo and Goldberg, D. E. (Eds.) (2010). *Philosophy and Engineering. An Emerging Agenda*. New York, Springer. 33

[7] The number of papers on the philosophy of engineering at ASEE conferences have slightly increased since 2013, but their still miniscule number is not my primary point. The primary point is that these papers uniformly see the philosophy of engineering as being different than and separate from the philosophy of engineering education. 33

[8] Morrison, E. E. (1977). *From Know-How to Nowhere: The Development of American Technology*. New York, New American Library, A Mentor Book Edition. 34

[9] Koen, B. V. (1985). *Definition of the Engineering Method*. Washington, DC, American Society for Engineering Education. 35

[10] Petroski, H. (1985). *To Engineer is Human: The Role of Failure in Successful Design*. New York, St. Martin's Press. 35

[11] Martin, M. and Schinzinger, R. (1983). *Ethics in Engineering*. New York, McGraw-Hill. 35

[12] Gravander, J. (1980). The origin and implications of engineers' obligations to the public welfare. *PSA: Proc. of the Biennial Meeting of the Philosophy of Science Association*, pages 443–455, Chicago, IL, The University of Chicago Press. 35

[13] James, W. (1896). *The Will to Believe and Other Essays in Popular Philosophy*. New York, Longmans, Green. 36

[14] *Loc. cit.* Ref. [2]. 37

[15] There are instances of these non-traditional types of learning experience that have been incorporated in four-year curricula; indeed, I have noted some of them elsewhere (see, for

example, [18]). However, I would make three points about these. First, even when non-traditional approaches are implemented and sustained within four-year curricula, students have insufficient exposure to them for the intended results to be achieved fully. Second, history has not been kind regarding the sustainability of such non-traditional approaches within four-year curricula. Third, I do not believe it is possible for these non-traditional approaches to become a general characteristic of four-year undergraduate engineering curricula. It would take another chapter or two to argue for these points, so suffice it to say they have been made by others over the years, both at annual ASEE conferences and elsewhere. 37

[16] Heywood, J. and Lyons, M. (2018). Technological literacy, engineering literacy, engineers, public officials, and the public. *ASEE Annual Conference Proceedings*. https://www.asee.org/public/conferences/106/papers/23287/view 37

[17] Korte, R., Mina, M., Frezza, S., and Nordquest, D. (2019). *Philosophy and Engineering Education: Practical Ways of Knowing*, vol. 2, Morgan & Claypool. 38

[18] Gravander, J., Slaton, A., and Neeley, K. (2002). Best practices for integrated curriculum design and administration: Objectives and exemplars. *Liberal Studies and the Integrated Engineering Education of ABET*, Report from the *National Science Foundation Funded Planning Conference at the University of Virginia*, pages 65–130, April 4–6, (Charlottesville: The University of Virginia, School of Engineering and Applied Science, 2003). 40

CHAPTER 4

Abstract Thought in Engineering Science: Theory and Design

Gregory Bassett and John Krupczak, Jr.

4.1 ABSTRACT

One goal of the philosophy of engineering is to more clearly distinguish engineering from science. This paper advances the suggestion that one distinction between the activities of science and engineering concerns the role of abstract thinking. A scientific theory unifies entities that are conceived of as existent in the world; an engineering design unifies existent entities with ones whose existence depends upon the design. The creation of scientific theory involves a single abstraction of a pattern that can unify existent entities. The creation of engineering design utilizes a double abstraction. An engineer must grasp an idea of purpose abstracted from any of its particular instantiations, and must also grasp the relation between this abstract idea of purpose and the existent entities, typically called components, relevant to the creation of the design. An implication for engineering education is an elevation of the study of components as functional elements in technological system design to be on par with current practice on analysis methods. In addition, engineers need familiarity with multiple paradigmatic examples of design patterns or system function structures so that they have resources for connecting conceptions of world and function.

4.2 INTRODUCTION

A well-developed philosophy of engineering should clarify its relationship to science. Such a clarification might help explain the connection between changes in technological design and scientific theories, the interdisciplinary research efforts of scientists and engineers, and the educational overlap of the two fields. In addition, articulating the differences between them might help resolve the perceived imbalance between the status attributed to scientific inquiry and that associated with the engineering design process [1, 2]. Such an articulation might also inform engineering education [3, 4].

The terms "science" and "engineering" each refer to a great range of practices, ideas, goals, and institutions. Nevertheless, it is common to group certain activities into one or the other, and such grouping does not seem completely arbitrary. This paper is an attempt to explain such grouping by articulating some distinctive features of each type of activity, specifically with regard to the role of abstract thought.

4.3   THE STANDARD ACCOUNT

Since at least Aristotle there has been a train of thought that distinguishes theoretical from practical knowledge, with the latter being an application of the former, and thus both posterior to it and of lower epistemic status [5]. These ideas have informed a common understanding of the difference between (theoretical) science and (practical) engineering. Thus, a common way to characterize the two is that science creates theories about the natural world while engineering creates artifacts to provide for human needs and wants [6–9]. This way of distinguishing between the two fields can be seen in popular discourse, academic literature [10], the division between funding "basic" or "applied" research [11], and the way that academic departments are named [12].

This standard account of the relationship between science and engineering has several flaws. It neglects the extent to which engineering involves much more than the application of scientific theories. It ignores the ways in which science often depends upon (and is thus not prior to) engineered technology. It is difficult to reconcile with engineering success often providing the reasons why scientific theories should be believed. By focusing only on products rather than methods, it leaves differences in the activities and education required for each unclear, and offers little reason to think that the two fields share much in common. The nature of those products also needs to be clarified. If "artifacts" refers only to physical objects, it seems clear that engineering does not create artifacts. After all, customarily civil engineers do not pour concrete and mechanical engineers do not machine engine blocks. If, on the other hand, "artifacts" can refer to intellectual products, then scientists could also be described as creating artifacts to provide for human needs and wants.

There may be reasons to discard the standard account entirely. However, the goal of this paper is not to overturn the standard account, but to ameliorate it, focusing specifically on the role of abstract thought in each field. We will assume that "science" and "engineering" refer to a coherent grouping of activities; that the knowledge gained from those activities is in some sense theoretical and practical, respectively; and that those activities are loosely, but not exclusively, connected to the disciplines that receive those names. By examining the products and methods of each field, suggestions might emerge about how they can be better distinguished from one another.

4.4   PRODUCTS

We are assuming that the practice of science produces theories. It might be that scientific theories are produced for a further purpose. It is also undoubtedly the case that science produces more than theories; other products of science include research programs, predictions, explanations of particular events, controversies, practical guidance, and objects of aesthetic appreciation. Theoretical production will be assumed here to be important and necessary for science, but not sufficient for it. In addition, with one exception, this paper is not an attempt to distinguish scientific theories from non-scientific or pseudo-scientific theories; thus, the broad category of theory-producing fields may need further analysis in order to separate science from non-science. We are assuming merely that theory production is at least partially definitive of science and not of engineering. The assumption that science produces theories does not entail that all individual scientists do so. Even those who do are usually clarifying, refining, or extending a theory rather than creating it whole. Individual scientists may perform a variety of activities that are not the production of new theories; however, these tasks can qualify as scientific only if they are a constituent part or consequence of activity that creates theories.

Theories are abstract and universal; they are not spatial or temporal entities (though expressions *of* them may be), and they are not about particular objects, but about the patterns instantiated within such objects.

What does the practice of engineering produce? As noted before, engineers are often not the proximal creators of artifacts, though they are involved with the creation of certain types of artifacts, namely those created to fulfill a function. The creation of artifacts to perform a known function can be called "craft." Craft, unlike accidental or perhaps artistic production of artifacts, requires a distinction between planning and execution. These two elements of craft exist temporally in that order: the craftsperson must know what is to be made, must grasp it as an object of thought, before she makes it. Thus, the planning must be prior to the execution, and must direct it. Engineering produces plans to be executed. However, the forming of plans is not enough. Everybody forms plans, often unconsciously or without being able to articulate them, and yet in doing so could only in the most remote sense be called an engineer. What is distinctive about an engineer's plans is that they are put into a communicable form. The plan has to be separable enough from its execution that the planner and executioner could be two different people. An engineer creates plans communicated through shared language (including words, gestures, models, sketches, specifications, technical drawings…). These communicated plans are more familiarly called "designs." Drawing and specification standards exist for the purpose of insuring that fabrication can be carried out without the intervention of the designer. The necessity of this condition is especially evident when considering that most modern products are developed by sizable design teams working in collaboration rather than individual engineers working alone. As with science and the production of theories, the notion that engineering produces designs does not mean that all engineering activity is such production or that all individual engineers design. Engineers perform a wide variety of activities; however, those activities can qualify as engineering only if they are a constituent part or consequence of activity that creates designs.

Designs, like theories, are abstract and universal: they are not spatial or temporal entities (though expressions of them are), and they are not particular objects but are rather the patterns to be instantiated within such objects.

Both theory and design are abstract universals and are attempts to solve a perceived problem. We suggest the difference between them can be grasped in terms of the problems they confront. Awareness of a problem occurs when there is a perceived lack of fit between one's ideas. Both theory and design aim to rectify the lack of fit by unifying previously disparate or conflicting entities into a coherent workable whole. The difference between them therefore lies in the type of entities that need to be unified.

The entities that a scientific theory unifies can be left relatively open: events, objects, kinds, observations, phenomena, experiences, conceptual definitions, and other theories may all be appropriate candidates. The requisite quality of them for present purposes is that they are conceived of as existent. This supposition of existence does not entail that such entities exist in the present; they could be past observations, future predictions, or objects enduring over time. It merely indicates that they are treated as fixed independent of the theory. It might be that what a theory unifies are post-theoretical entities, but if so those post-theoretical entities are treated as stable objects that the theory also unifies. The claim that theory unifies disparate entities does not imply that the theory is known posterior to them. The question of the epistemic priority of theory or observation is not one we are addressing.

In contrast, design does not unify disparate *existent* entities, but rather unifies existent entities with ones whose existence depends upon the design. Part of the problem to be solved is, like science, regarded as about the world. The world is often conceived of differently by an engineer and scientist—in fact, the world is often conceived of differently from engineer to engineer [13, 14]. However, in both fields the world is taken as an existent given. Design, unlike theory, must unify that conception of the world with a conception of a to-be-fulfilled purpose, known as a "function" when referring to the artifact designed. The fulfillment of purpose is not regarded as independently existent, but is rather regarded as something that exists in the world only as a consequence of the design. The usual purpose of refrigerator, for example, is to reduce the temperature of food to retard spoilage. When the refrigerator is being designed, this reduction of temperature is conceived of as existing only after, and as a consequence of, the existence of the design. Most modern technologies such as aircraft, instantaneous global telecommunications, personal automobiles, and computerized storage and retrieval of information exist only as a consequence of being designed to fulfil a specific purpose.

Half of a design must therefore be conceived of as a conditional future, and thus the design must be thought of as about an addition to the world. In contrast, the solution to a scientific problem is conceived of as internal to—already existing in—the world. The object of theory is not understood as something external to the experienced world, but as a property intrinsic to it.

Thus, although theories may be functional, they are not conceived of as functional in science. In contrast, design is understood as something external to the experienced world. Designs are therefore seen as productive of artifacts; theory is not similarly productive because its object is conceived of as already existent.

4.5   METHODS

In addition to their products, science and engineering can be distinguished by their methods, specifically in the sort of abstraction needed in each. It may be the case that all perceptions of the world require a degree of abstraction. Mere sensory input cannot be a basis for knowledge until it has been organized into a perceptual experience, with a distinction between foreground and background, object and context. It is this organizing, perhaps an organization requiring prior concepts, that allows for perception by isolating a stable object—the object of perception—from the flux of sensation, and by requiring input from the subject, allowing the thought to be attributed to a mind. For the most part these sorts of abstraction are not relevant here, not because they do not exist or are not part of science and engineering, but only because any abstraction necessary for perception is assumed to be common to both fields. Similarly, any conception of what it means to be coherent, true, or functional may be a requisite abstract idea for both fields, but will not be examined here. A scientist must have in mind a conception of knowledge in order to create good theory. An engineer must have a conception of functionality, not merely of a function, in order to create good design. In both cases, the conception of what it means to create good output is considered intrinsic to the process rather than as a goal to be accomplished from it.

Scientific theories are developed and tested by means of a set of accepted processes, stated and unstated, as applied by trained practitioners. It is not the intention here to examine the rationality or reliability of those processes. Instead, we are only focused on the type of abstract thought necessary for theory production. A theory understood only in terms of certain paradigmatic examples, rather than being abstracted from them, would be unable to explain new experimental results. Being new, the results would not be the same as those examples, and thus could not be recognized as instantiations of the theory. Thus, a scientist must be able to understand a theory abstracted from its particular instantiations in order to recognize new experimental results. However, a scientist must also be able to perceive and imagine particular instantiations of the theory in order to design an experiment and see how it is relevant to particular experiences. A well-known and controversial example of how scientists imagine particular instantiations of an abstract theory and design experiments accordingly is provided by Eddington's 1919 experiments to measure deflection of light by the sun, as predicted by Einstein's general theory of relativity [15, 16]. Presently, in the quest for grand unified theories in physics, the issue of what measurable phenomena a particular theory might encompass is an important question and a source of controversy among practitioners in the field [17].

The suggestion here is that scientific theory creation requires relating and uniting only existent entities. Thus, excluding any abstraction necessary to be aware of those existent entities, the production of scientific theory requires a single abstraction. In contrast, design production requires a double abstraction. An engineer must grasp an idea of purpose abstracted from any of its particular instantiations prior to creating a design that will achieve it. In the absence of a prior abstract conception of purpose, the engineer would be unable to know whether any new artifact designed was functional. An engineer must also grasp the relation between this abstract idea of purpose, conceived of as not an existent given, and the existent entities, typically called components, relevant to the creation of the design. This stage of abstraction can be complex: for instance, insofar as the function of an artifact is dependent on a relationship between subfunctions, an engineer must be able to grasp these relations, so that the inputs and outputs of each subfunction work together coherently. In other words, an engineer must have a conception of the relation between function and structure, abstracted from particular instantiations of it, in order to create a new design.

As an example consider a portion of an automobile assembly line such as painting. The manufacturing engineer must first grasp the abstract purpose of painting the car body independent of any particular instance of it. The second abstraction is consideration of the available functions provided by existing components that may provide elements of the design. These may include spray nozzles to distribute paint, tubing to transport paint, tanks to pressurize paint, along with components such as switches, timers, color sensors, and ventilation fans. Each of these components has a function which the designer might choose to employ to achieve the overall purpose of painting the car. The engineer must also grasp the necessary inputs and specific outputs of each component to envision how a particular component may be interconnected with others to form a complete system.

Designing a photovoltaic power system for a home also illustrates the double abstraction. The engineer must first comprehend the purpose of the intended system to supply electrical power. The engineer must then grasp the relation between this abstract idea and the abstract functional capabilities of existent entities relevant to the creation of the design. Such capabilities would include photovoltaic modules providing the function of converting sunlight to electrical energy, conductors supplying a likely required function of transporting electrical energy, switches actuating flow of current, and meters measuring relevant characteristics such as voltage and current. The engineer must also comprehend the needed inputs and provided outputs of each device to envision how each subfunction interacts with other potential components.

It is only after those abstract ideas of function and function-world relationship are thought of together that a design that would connect them can be imagined. This double abstraction does not necessarily make the production of design more difficult or complicated than the production of theory, but it does entail that it is of a different character.

How is a design imagined once the abstract ideas are in place? It is easy at this stage of the process to wave one's hands about creativity or a non-intellectual engineering skill, much as the source of a particular jazz improvisation may remain a mystery. Some of that may be unavoidable. However, like any good jazz improviser, an engineer will need to have practiced common patterns of designs in order to develop new ones that resemble those patterns. In other words, an engineer must have worked with a number of particular designs and abstracted out of them a pattern which can serve as the basis for a new design. Thus, in order to create a design, an engineer must have not only engaged in abstract thought to grasp the function, and to understand how the function relates to the world, but must have a set of abstracted design patterns to be imaginatively tested and manipulated to find a fit with that relation. For example, designing the photovoltaic power system is facilitated if the engineer is already familiar with commonly used patterns by which existing components have been assembled into photovoltaic systems. The designer can also draw on more generalized patterns or function structures employed in electrical systems. As with the production of designs, producing theories requires someone who has familiarity with common patterns of theory; thus, a scientist must have worked with a number of other theories and recognized an abstract pattern within them that could be instantiated with a new set of existent entities.

It is a misleading commonplace to say that engineering is applied science. However, there is a grain of truth to it: engineers rely upon scientific theory to form their conceptions of the existent world and to frame an idea of possible futures; those ideas make up half of the relational idea they have to form in order to create a design. Nevertheless, the scientific ideas that engineers use have to be transformed by them so that they can fit into a coherent relation with the function. In other words, engineers do not simply apply ideas ready-made by science; they transform them so that they can occupy a place in the domain of functionality. Analogously, physicists often have to transform the ideas of mathematics in order for them to be relevant to a world of matter [18].

It would also be misleading to characterize science as merely generalized engineering success. However, there is also a grain of truth to that: even outside of the technological infrastructure that scientists depend upon, science relies upon designed products (i.e., experimental results) to generate the data that science requires. What distinguishes science from philosophy or mathematics is often thought to be that it is *less* abstract—its products (i.e., its theories) are instantiated more directly in experience, and thus have a more experimental aspect to them. This conception of science may be flawed, but assuming there is some truth to it, suggests that what distinguishes science from mathematics and philosophy is that science relies more directly upon engineering. Scientists use *designed* experiments. Experiments are designed to get a predicted result, and whether or not that design succeeds is considered relevant for the acceptance of theory. Thus, science relies on engineering in designing experiments to produce a functional expected result.

Given this dependence between the two fields, it should be no surprise that scientific and technological progression often occur jointly. Theories are sometimes created to fit prior design successes, such as the development of optics to fit the success of the telescope; but design success also relies upon implicit or explicit acceptance of approximately accurate theory.

4.6   EDUCATION

Thus far, we have been attempting to fill in some gaps in a standard account of the distinction between engineering and science. We are proposing that, in this standard account, both engineering and science create communicable entities that are not by themselves instantiated in the world, but are rather patterns abstracted from such instantiations. However, the creations differ: science creates theories, which fit together two or more ideas about the existent world; engineering creates designs, which fit together an idea about the existent world and an idea about function. This difference in what is produced requires differences in methods, and we have focused on the role of abstract thought in those methods. Theory creation requires a single abstraction of a pattern in such a way that those patterns could by design (i.e., through experiment) be re-instantiated in particular observations. Design creation requires the abstract ideas of function and of function-structure relationship.

Assuming these differences in outputs and methods, what sort of significance does this have for education? In particular, what skills do engineers need that would differentiate their education from scientists', and can these ideas help explain some of the educational practices of the two fields?

As was mentioned earlier, engineering and science rely upon each other. Without an approximately accurate idea of how the world works, an engineer's designs will be unsuccessful, since one part of the relation an engineer must grasp will be flawed. In addition, without the ability to frame that idea in communicable language, it is possible to be a craftsperson but not to create designs that could be executed by another—in other words, it is impossible to be an engineer. Thus, engineering relies upon science for a communicable, approximately accurate conception of the world [19]. Similarly, scientists need to have some practice in engineering design. Science without engineering is at best very poor philosophy. Scientists, or at least some scientists, need to grasp the relationship between conceptions of the world and how they are realized by designed experiment.

It is in the understanding of function that engineering is distinguished from science, and thus successful engineering education requires not merely an approximately accurate picture of the world, but discussion about, and practice with, the notion of function: for example, how a single function can have multiple possible instantiations, how to understand what the best balance between multiple functions is, and how function and structure are related. A capability in appreciating and envisioning multiple possible forms for achieving a particular function is central to the education of creative and innovative engineers. A simple example of such a capability would involve the recognition that the function of conducting electrical current typically accomplished using copper wire can also be achieved through use of other metals, conductive polymers, or liquids containing ions. A more complex example would involve the understanding that an electrical signal varying between two voltage values at a specific rate (i.e., a square wave) can be achieved by construction of an appropriate circuit using discrete components, utilization of an application-specific integrated circuit, or appropriate programming of a microprocessor.

An example of how engineers need to balance multiple functions is the trade-offs an automobile designer must be aware of between speed and fuel economy, or safety and cost.

In addition, engineers need practice with common components and design patterns, so that they have resources for connecting conceptions of world and function. Thus, a study of multiple paradigmatic examples of engineering excellence and an appreciation of engineering history can be beneficial. Engineering education has emphasized analysis of particular, well-defined, physical situations such as the RC circuit, simply supported beams, and internal flows of Newtonian fluids. Recently, the importance of design as a process has been recognized and the study of design methodologies has helped in development of student competence in the first abstraction that is the conception of purpose abstracted from any of its particular instantiations. To date, the second abstraction, grasping the relation between abstract purpose and existent entities, has been underdeveloped in formal engineering education. Simple components such as resistors, capacitors, pulleys, and gears do routinely appear in analysis problems. However, development of facility in both abstractions required in engineering design calls for a diverse and nuanced inclusion of more sophisticated components such as motors, pumps, heat exchangers, dampers, engines, and batteries in the context of their appearing as functional elements in specific types of technological systems.

The standard account of differences between science and engineering outlined here also seems to fit with social sciences and their engineering counterparts. For instance, political science, psychology, and economics commonly recognize themselves as having aspects of both theoretical science and practical engineering. The sketch offered above might help for understanding the relationship between those aspects as well.

4.7   NOTES AND REFERENCES

[1] Heywood, J. (2011). A historical overview of recent developments in the search for a philosophy of engineering education. *41st ASEE/IEEE Frontiers in Education Conference*. Rapid City, South Dakota, October 12–15. 41

[2] Michfelder, D. P., McCarthy, N., and Goldberg, D. E. (Eds.) (2013). *Philosophy and Engineering: On Practice, Principles, and Process*. Dordrecht, Springer. 41

[3] Heywood, J., McGrann, R., and Smith, K. A. (2008). Continuing the FIE2007 Conversation on: Can philosophy of engineering education improve the practice of engineering education? *38th ASEE/IEEE Frontiers in Education Conference*. Saratoga Springs, New York, October 22–25. 41

[4] Heywood, J. (2008). Screening curriculum aims and objectives using the philosophy of education. *38th ASEE/IEEE Frontiers in Education Conference*. Saratoga Springs, New York, October 22–25. 41

[5] See, e.g., Aristotle, *Metaphysics 1.1*, or *Nicomachean Ethics*, 6:1–7. 42

[6] Changing the conversation: Messages for improving public understanding of engineering, (2008). *Committee on Public Understanding of Engineering Messages, National Academy of Engineering*. Washington DC, National Academies Press. 42

[7] Pearson, G. and Young, A. T. (Eds.) (2002). *Technically Speaking: Why All Americans Need to Know More About Technology*. Washington, DC, National Academies Press. 42

[8] Adams, J. L. and James, L. (1991). *Flying Buttresses, Entropy, and O-Rings: The World of an Engineer*. Cambridge, MA, Harvard University Press. 42

[9] Billington, D. and Billington, D. Jr. (2006). *Power, Speed, and Form: Engineers and the Making of the Twentieth Century*. Princeton, NJ, Princeton University Press. 42

[10] See, e.g., Feibleman, J. K., Pure science, applied science, and technology: An attempt at definitions, in Mitcham. C. and Mackey, R. (Eds.), *Philosophy and Technology*, pages 33–41, New York, The Free Press. 42

[11] Pitt, J. C. (2000). *Thinking About Technology*, p. 2 ftnt. p. 1, New York, Seven Bridges Press. 42

[12] For example, engineering departments are often called departments of applied science. For further examples of ways that this standard account appears as an assumption in public discourse, see, Goldman, S. (2004). Why we need a philosophy of engineering: A work in progress. *Interdisciplinary Science Reviews*, 2(2):164–166. 42

[13] Vincenti, W. (1990). *What Engineers Know and How They Know It*, p. 5, Baltimore, MD, Johns Hopkins University Press. 44, 51

[14] *Ibid.* pages 7–8. Vincenti draws a distinction between "normal design," and "radical design." One of the differences he proposes between them is that what is taken as given is more extensive and in the former than the latter. 44

[15] Einstein, Albert (1916). The foundation of the general theory of relativity. *Annalender Physik*, 49(7):769–822. 45

[16] Dyson, F. W., Eddington, A. S., and Davidson, C. (1920). A determination of the deflection of light by the Sun's gravitational field, from observations made at the total eclipse of 29 May 1919. *Philosophical Transactions of the Royal Society*, 220A:291–333. 45

[17] For example in chronicalling the controversy surrounding string theories, physicists Smolin and Woit emphasize the centrality of testable hypotheses and predictions that can be experimentally verified. Smolin, L. (2006). *The Trouble With Physics: The Rise of String Theory, the Fall of a Science, and What Comes Next*. Boston, MA, Houghton Mifflin Harcourt. Also Woit, P. (2006). *Not Even Wrong: The Failure of String Theory and the Search for Unity in Physical Law*. New York, Basic Books. 45

[18] See, e.g., Aristotle, *Physics 2.2* and Goldman (2004), 166. 47

[19] *Loc. cit.* Ref. [13], pages 207–222 for a more detailed examination of the types of knowledge of the world required for design. 48

Authors' Biographies

JOHN HEYWOOD

John Heywood is a Professorial Fellow Emeritus of Trinity College Dublin His primary interest is in education for the professions, especially engineering, management and teacher education. He was awarded the best research publication award of the division for the professions of the American Educational Research Association in 2006 for his book "Engineering Education: Research and Development in Curriculum and Instruction" published by IEEE/Wiley. His other publications include "Learning, Adaptability and Change: The challenge for education and Industry", "The Human Side of Engineering", "Empowering Professional teaching in Engineering; Sustaining the Scholarship of Teaching". He was a co-author of "Analysing Jobs".

WILLIAM GRIMSON

William Grimson received his B.A. and B.A.I. from Trinity College Dublin and his M.Sc. from the University of Toronto. He is a Charted Engineer and a Fellow of Engineers Ireland of which he has been President. Now retired. He worked as a Research and development engineer for Ferranti Ltd before joining the academic staff of the Dublin Institute of Technology. His academic output was and remains eclectic, ranging from publications in areas as diverse as plasma physics, clinical information systems, philosophy of engineering, and development issues.

JERRY W. GRAVANDER

Jerry W. Gravander is Associate Dean of the School of Arts and Sciences and Distinguished Service Professor of Philosophy in the Department of Humanities and Social Sciences at Clarkson University. He has undergraduate degrees in chemistry from Illinois Institute of Technology and in philosophy from the University of Tennessee, Knoxville, and a Ph.D. in the history and philosophy of science from the University of Texas at Austin. He has written and presented widely on liberal education for engineering students, as well as engineering ethics and the impacts of science and technology on society. He was the 1996 recipient of the Sterling Olmsted Award of the American Society for Engineering Education's Liberal Education Division.

GREGORY BASSETT

Gregory Bassett is a lecturer in philosophy at Hope College in Holland, Michigan. He received a Ph.D. in Philosophy from Indiana University, a Master of Music degree from the New England Conservatory, and a B.A in Philosophy from Swarthmore College. Dr. Bassett's research interests are primarily focused on ethics and action theory, including topics such as weakness of will and the relationship of desire and action.

JOHN KRUPCZAK, JR.

John Krupczak, Jr. is a professor of engineering at Hope College in Holland, Michigan. He has been a Senior Fellow of the Center for the Advancement of the Scholarship of Engineering Education (CASEE) of the National Academy of Engineering. Krupczak was founding chair of the Technological and Engineering Literacy Division of the American Society for Engineering Education (ASEE). From 2013–2016 he served as a Program Director in the Division of Undergraduate Education at the National Science Foundation. Krupczak received a Ph.D. in Mechanical Engineering from the University of Massachusetts and a B.A. in Physics from Williams College.

Printed in the United States
by Baker & Taylor Publisher Services